Sustainable Textiles: Production, Processing, Manufacturing & Chemistry

Series Editor

Subramanian Senthilkannan Muthu, Chief Sustainability Officer
Green Story Inc. Tung Chung, Hong Kong

This series aims to address all issues related to sustainability through the lifecycles of textiles from manufacturing to consumer behavior through sustainable disposal. Potential topics include but are not limited to: Environmental Footprints of Textile manufacturing; Environmental Life Cycle Assessment of Textile production; Environmental impact models of Textiles and Clothing Supply Chain; Clothing Supply Chain Sustainability; Carbon, energy and water footprints of textile products and in the clothing manufacturing chain; Functional life and reusability of textile products; Biodegradable textile products and the assessment of biodegradability; Waste management in textile industry; Pollution abatement in textile sector; Recycled textile materials and the evaluation of recycling; Consumer behavior in Sustainable Textiles; Eco-design in Clothing & Apparels; Sustainable polymers & fibers in Textiles; Sustainable waste water treatments in Textile manufacturing; Sustainable Textile Chemicals in Textile manufacturing. Innovative fibres, processes, methods and technologies for Sustainable textiles; Development of sustainable, eco-friendly textile products and processes; Environmental standards for textile industry; Modelling of environmental impacts of textile products; Green Chemistry, clean technology and their applications to textiles and clothing sector; Eco-production of Apparels, Energy and Water Efficient textiles. Sustainable Smart textiles & polymers, Sustainable Nano fibers and Textiles; Sustainable Innovations in Textile Chemistry & Manufacturing; Circular Economy, Advances in Sustainable Textiles Manufacturing; Sustainable Luxury & Craftsmanship; Zero Waste Textiles.

Sadhna • Rajesh Kumar • S. Greeshma
Editors

The Art and Craft of Modern Textile Design

Woven Whimsy

Springer

Editors
Sadhna
Department of Fashion Design
Woxsen University
Hyderabad, Telangana, India

Rajesh Kumar
Department of Fashion Design
Woxsen University
Hyderabad, Telangana, India

S. Greeshma
Department of Fashion Design
Woxsen University
Hyderabad, Telangana, India

ISSN 2662-7108 ISSN 2662-7116 (electronic)
Sustainable Textiles: Production, Processing, Manufacturing & Chemistry
ISBN 978-3-031-86796-5 ISBN 978-3-031-86797-2 (eBook)
https://doi.org/10.1007/978-3-031-86797-2

This Springer imprint is published by the registered company Springer Nature Switzerland AG
The registered company address is: Gewerbestrasse 11, 6330 Cham, Switzerland

Preface

Textile design, as a discipline and practice, sits at the intersection of art, craft, culture, and technology. Over the centuries, textiles have served as both functional objects and artistic expressions, embodying the narratives of societies, their technologies' advancements, and their makers' creativity. In the contemporary era, textile design has evolved to address the complexities of modern life, engaging with issues such as sustainability, cultural preservation, and technological innovation. *The Art and Craft of Modern Textile Design* is a timely and significant contribution to this evolving field, offering a multidisciplinary perspective that bridges the traditional with the contemporary and the local with the global.

This edited volume is a testament to textile design's dynamic and ever-changing nature. It brings together a diverse group of contributors, including academics, practitioners, and industry experts, whose work explores the multiple dimensions of textiles as both material and medium. The book delves into the rich traditions of textile crafts, emphasizing their cultural and historical significance while highlighting the innovative approaches reshaping the discipline in the twenty-first century.

One of the central themes of this volume is the interplay between tradition and innovation. The contributors have shed light on how traditional textile practices, rooted in centuries-old techniques, continue to inspire and inform contemporary design. These practices, often tied to specific cultural and geographic contexts, offer a wealth of knowledge and creativity that can guide modern designers in pursuing sustainable and meaningful solutions. At the same time, the book explores how advancements in technology—from digital printing to smart textiles—are expanding the boundaries of what is possible in textile design, opening up new opportunities for creativity and functionality.

Sustainability is another critical focus of this volume, reflecting the urgent need for the textile industry to address its environmental and social impacts. The contributors engage with pressing issues such as resource conservation, waste reduction, and ethical labour practices, offering insights into how textile design can contribute to a more sustainable future. By integrating sustainable principles into their work, designers and industry professionals are not only reducing the ecological footprint of textiles but also redefining their value and significance in a rapidly changing world.

In addition to its thematic richness, *The Art and Craft of Modern Textile Design* stands out for its interdisciplinary approach. The chapters draw on perspectives from art, design, anthropology, history, and engineering, among other fields, creating a holistic understanding of textiles as both material and a cultural artefact. This interdisciplinary lens is particularly valuable in today's interconnected world, where the challenges and opportunities of textile design require collaborative and cross-disciplinary solutions.

This book also underscores the importance of storytelling in textile design. Textiles are often deeply personal and intimate objects, carrying the stories of their makers, users, and communities. Through their exploration of various case studies and practices, the contributors reveal how textiles can serve as powerful mediums for storytelling, preserving cultural heritage, and fostering social connection.

As the global textile industry continues to evolve, *The Art and Craft of Modern Textile Design* serves as both a reflection of current trends and a guide for future directions. It challenges readers to think critically about the role of textiles in society and to consider how design can address the complex challenges of our time. By bridging the past and the future, the traditional and the innovative, this book provides a comprehensive and thought-provoking exploration of the art and craft of textile design.

This volume will be a valuable resource for scholars, students, practitioners, and anyone interested in textiles and design. It not only deepens our understanding of the field but also inspires us to envision new possibilities for the future of textile design. I commend the editors and contributors for their outstanding work and for creating an intellectually rigorous and engaging book.

Hyderabad, Telangana, India Sadhna

Acknowledgments

It is with profound gratitude that I acknowledge the contributions of all those who played an integral role in the development of *The Art and Craft of Modern Textile Design*. This edited book represents a culmination of collaborative efforts, intellectual engagement, and a shared commitment to advancing the fields of textiles, art, and design innovation.

I extend my sincere gratitude to the editorial team, Rajesh Kumar and Greeshma S, whose steadfast support and intellectual guidance have been indispensable throughout this process. Their dedication and expertise have provided both inspiration and a foundation for refining this volume.

Special acknowledgment is due to the reviewers who generously offered their time and scholarly expertise. Their insightful critiques and constructive feedback have been instrumental in ensuring this work's academic rigour and comprehensiveness. The high quality of this publication owes much to their careful evaluation and thoughtful recommendations.

I am deeply grateful to the team at Springer for their unwavering professionalism and commitment to excellence. Their expertise in academic publishing, coupled with their alignment with the themes of quality and sustainability, has significantly contributed to the successful realization of this project.

Furthermore, I would like to recognize the invaluable contributions of textile and garment industry professionals, artisans, researchers, and policymakers, whose work continues to enrich the discourse on textile design and sustainable practices. Their ongoing efforts and dedication have been pivotal in shaping the themes and perspectives presented in this book.

This volume is also an honour to the global community of scholars, practitioners, and readers who are dedicated to exploring sustainable practices and the evolving role of textiles and design in society. I extend my deepest gratitude to the readers for engaging with this work and contributing to the broader dialogue surrounding sustainability and innovation in textile design.

Finally, this book is dedicated to all individuals and institutions aspiring to advance knowledge, encourage innovation, and promote sustainable development within the textile and garment industries. It is my hope that *The Art and Craft of Modern Textile Design* serves as a meaningful academic contribution to the field and inspires further scholarship and practice in this essential area.

Sadhna

Contents

Introduction to Modern Textile Design: Evolution, Techniques, Technology, and Sustainability

Sadhna, S. Greeshma, Rajesh Kumar, and E. Mokanaasri

1 Introduction

Modern textile design develops fabrics with creativity, state-of-the-art technical methods, current aesthetics, and practical functions. This aspect depicts the shift from ordinary designs based on tradition or ornamentation to more contemporary minimalist or practical creations. Modern design satisfies aesthetic functions and serves the useful functionalities of the fabric. Such concepts involve thinking of and making designs according to current trends and technologies available. The methods of this could include various forms of weaving, knitting, printing, and fabric manipulation processes. This approach usually aims to emphasize sustainability, adaptability, and innovation in the choice of materials and production methods, reflecting the needs of society [1].

Textile design significantly affects daily life through personal identity and cultural expression. It is a function and a medium for telling stories and representing cultures. Designers get inspiration from their cultural background, using patterns, colours, and materials familiar to specific communities. Cultural integration enriches the design process and allows for diverse expressions in modern fashion and home decor [2].

Furthermore, textile design represents and leads cultural trends. For example, the popularity of vintage patterns and the move towards sustainable practices in textile design parallel societal trends toward nostalgia and greater environmental awareness. Increasing awareness of consumer choices further expands the market

Sadhna · S. Greeshma (✉) · R. Kumar · E. Mokanaasri
Department of Fashion Design, School of Arts and Design, Woxsen University, Hyderabad, Telangana, India
e-mail: dr.sadhna345@gmail.com; greeshma2551997@gmail.com; kumar.rajeshrk2030@gmail.com; mokanaadevreddy.98@gmail.com

Sadhna et al. (eds.), *The Art and Craft of Modern Textile Design*, Sustainable Textiles: Production, Processing, Manufacturing & Chemistry,
https://doi.org/10.1007/978-3-031-86797-2_1

for unique, responsibly produced textiles, which leads to trends and directions in market demand.

Artistic expression and social commentary are intertwined with textile design to evolve into a great heritage symbol of innovation. Based on history tracing the Indus Valley Civilization, textiles are used for practical needs, storytelling, traditional techniques, and modern techniques to create diverse designs, reflect stories, and build communities. Some patterns carry deep meanings; they reveal values, beliefs, and histories. The symbolic colouration and designs of African tribal textiles tell stories. That said, this connection calls for conserving tradition while adapting it to present-day contexts to ensure textiles continue to be a rich form of cultural expression and innovation [3].

This influence of globalization makes a blend of various cultural practices possible in this modern design, where such diversity in textile designs and patterns has been invented to celebrate diversities. With this inspiration coming from many sources, today's designers are coming out with hybrid patterns that could be seen as traditional motives and modern aesthetics altogether, making the textile industries richer and cross-culturally more communicative of artistic expressions. The increased use of digital printing and other high-tech manufacturing techniques has even more democratized textile design, so artists can make designs with a level of intricacy that would be difficult to produce by hand [4].

Textiles are increasingly being used as social commentary, discussing issues such as sustainability and ethical labour. Designers use organic, eco-friendly materials and combine traditional designs with modern awareness. This change helps support local economies, preserves cultural heritage, and fosters environmental consciousness. Innovating responsibly, this textile design honours its cultural roots while contributing to a sustainable and ethically aware future.

Textiles play a huge role in personal identity and expression. The reason is that the world constantly strives to do something novel to express its identity. Therefore, handmade textiles have become a medium through which people can link up their history or create a new culture through stories. The fabrics' use of colour and texture can be associated with emotions and mark personal stories, thus an intense way to express personal identity through them [3]. In this context, the element of flexibility in textile design is the dynamic aspect it takes by changing with time.

2 The Evolution of Textile Design

Textile design history spans thousands of years and reflects the ingenuity of humans in the face of cultural diversity. Ancient civilizations used early hand-weaving to develop intricate patterns and symbolic meanings. Mechanization during the Industrial Revolution dramatically increased the production and variety of such products. Digital printing and sustainability have recently revolutionized the field, reflecting societal changes and technological developments. Evolution thus underlines the fact that textiles have been playing a role in expressing identity and adapting to changing times.

Textiles can trace ancient civilizations; evidence suggests that the first of these fabrics was produced in 5000 BCE [5]. Cotton and silk are still woven in their respective textiles in India and China. Ancient Egyptian linen fabrics were produced in the range of 4000–5000 years ago [6]. Its use of vegetable dyes gave wool and cotton a vibrant colour. Similarly, in China, the production of silk dates back to about 3630 BCE and therefore, high-quality damasks emerged-the woven fabric containing complex designs, which gained immense demand [7].

With the emergence of more advanced civilizations and the expansion of trade routes, textiles were a precious good. Early in the Han Dynasty, around 114 BCE, the Silk Road was established to allow luxury textiles of the East to cross into the West. This network not only spread the existence of weaving techniques and materials across cultures but also enriched textile design worldwide. For instance, damask weaving is said to have originated in China but indeed flourished in Italy in the fourteenth century when European craftsmen adopted and adapted those techniques [8].

During the Middle Ages, the evolution of textiles developed significantly. Improving the dyeing technique and creating new European markets provided various colours and patterns. Moreover, guilds related to the production of textiles have started to emerge during this era, which standardized their practices and improved quality control [9]. Woollen textile's Introduction into England further diversified the market since artisans experimented with other fibres and weaving methods.

The Industrial Revolution brought significant changes to textile design. Spinning jenny (Fig. 1) and power loom introduced excellent efficiency into fabric production while bringing the cost of labour down, thus making textile production more accessible and inspiring innovations in fashion. Synthetic fibres such as nylon and polyester also emerged during this period, increasing the options for designers and manufacturers [11].

The Arts and Crafts movement, which lasted until the late nineteenth century, opposed the industrialist approach by emphasizing the quality produced by hand and traditional techniques. Figures such as William Morris championed returning to simple, beautiful designs characterized by craftsmanship over mass production. Aesthetic values, along with functionality, paved the way for modern textile art [12].

During the twentieth century, textile design underwent tremendous transformations. The modernist art movements had led artists to experiment with abstract forms and geometry. From these explorations came diverse styles ranging from culture in its entirety [11]. Technology also played an essential part; CAD computer-aided design revolutionized conceptualization and production for textile designers. Tools such as Adobe Photoshop allowed these designs to be created digitally prior to printing on fabric. Contemporary textile design has also adopted sustainability as a guiding principle. Designers increasingly focus on eco-friendly materials and production methods to reduce environmental impact. Innovations such as digital fabric printing have made it possible to produce smaller runs of customized fabrics with minimal waste [13].

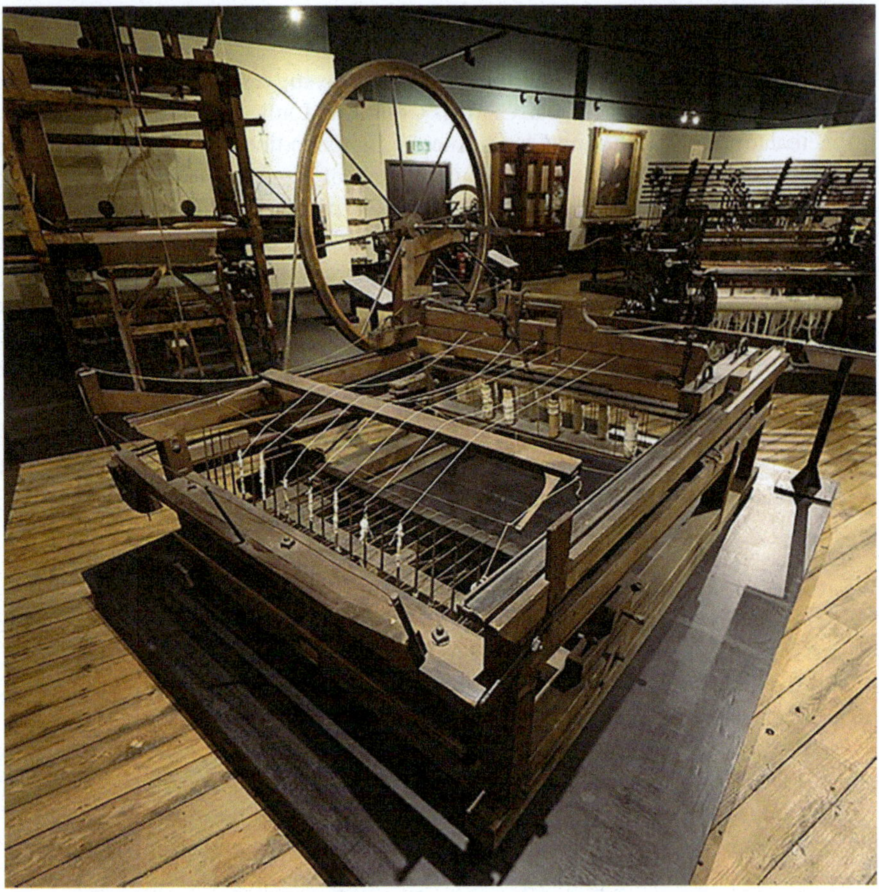

Fig. 1 Spinning Jenny at Blackburn Museum and Art Gallery [10]

3 The Role of Textile Designers

Textile designers play an essential role in the fashion and interior design industries. They often focus on fabric and pattern development for many different products. Their work shapes not only aesthetics but also functionality and sustainability in textile production.

The most fundamental part of a textile designer's job is creating innovative prints and patterns. This requires both creativity and technical skills since designers use a combination of traditional techniques like hand drawing and computer-aided design (CAD) software. Designs are vital for garments and home textiles since they form the visual identity of the products. A designer needs to be attuned to current trends so that their creations will resonate with the target audiences, which can be very different across demographics [14].

Besides aesthetics, textile designers also have to develop fabrics. For instance, they will choose the appropriate fibres, colours, and finishes to develop a particular type of fabric for certain requirements. Textile designers work with new materials or adapt existing ones to develop fabrics that can meet various performance characteristics, including durability, comfort, and environmental effects. This requires thorough knowledge of textile science and manufacturing methods.

Sustainability is the fashion industry's focus, and textile designers lead this trend of eco-friendly innovation. Their concern is sustainable materials and processes to minimize waste and fulfil consumers' demand for responsible products. Designers assess textiles regarding texture, weight, and cost; they balance client needs with practicality and production feasibility. They oversee production, ensure it is done in line with the original designs, coordinate fabric testing, and manage supplier relationships to maintain quality. This collaboration is very important in modern fashion design. Often, the materials involved define the result when working with a collection's designer, which is common in developing fashion products [14].

4 The Role of Technology in Textile Design

Technology in the textile industry has significantly changed over the last two decades. New developments have transformed production and given designers' imaginations scope. This essay examines the many aspects of the role technology plays in textile design, from digital technologies to smart textiles and sustainability to automation.

Technology has changed fashion dramatically. It affects design, style, and even the manufacture of clothes. From the Stone Age, when clothes met only basic needs, to modern synthetic fabrics, technology enabled the production of specialized garments for extreme conditions, such as space and deep-sea exploration. Mechanized production increased the output of textiles, while automation improved quality control. Other innovations include advanced fabrics, like moisture-wicking and dry-fit sportswear. Computers have revolutionized design through CAD software, streamlining the creative process. Future innovations, such as fabrics that adapt to environmental conditions, promise even greater possibilities, making technology integral to the textile and fashion industries [15].

Digital printing technologies have also changed the way patterns are applied to fabrics. Designers no longer have to rely on screens and all the setups that come with them; instead, inkjet technology allows for designs to be applied directly to textiles. This process speeds up production and allows for greater customization and personalization of fabrics. Designers can create complex patterns with vibrant colours and intricate details that were previously challenging to achieve [16].

3D printing is one of the most revolutionary technologies changing the way textile designs are done. This process allows for creating distinct fabric structures and components, which can be adjusted according to requirements. Brands are already using 3D printing to create footwear with soles that can be modified for enhanced performance while minimizing the loss of material. It also enables innovation in fashion through small batch production of highly detailed designs [17]. Figure 2 depicts the 3D printed textile.

The advent of smart textiles marks a giant leap forward in textile design. Such fabrics contain electronic parts that enable them to engage with the environment or wearer. For instance, some clothes with sensors can provide information on body temperature or heart rate, giving real-time information to athletes or health-concerned individuals. This technology integration in textiles boosts functionality, and new markets are found for health-monitoring apparel [19].

In addition, nanotechnology provides new properties for fabrics by giving them water resistance and stain repellency and enhancing their antimicrobial features. These can be done by playing with materials at the nanoscale to make the fabrics durable and functional for consumers' needs for comfort and hygiene. Therefore, innovations are highly applied in the sportswear and medical textiles sectors for performance and safety purposes [20].

As environmental concerns grow, the textile industry increasingly adopts sustainable practices made possible by technology. Digital technologies enable designers to create more efficient production processes that do not waste much material. For example, through digital fabric twins, a designer can test fabric properties digitally before their physical creation, thus requiring fewer samples and less waste material in the process [21].

Advancements in recycling technologies are now allowing for the efficient reuse of old textiles to make new products. Upcycling and biodegradable textiles are gaining popularity because consumers demand more sustainable options. Innovations in

Fig. 2 3D Printed outfit [18]

materials science have led to the development of bio-based fabrics that reduce reliance on non-renewable resources while maintaining quality [17].

Integrating sustainable practices is no longer a fad; it is the future as companies realize the need for corporate social responsibility. Through the use of technology, companies can reduce their environmental footprint while still catering to consumers' demands for quality products.

Automation has transformed the textile industry, enhancing precision and efficiency in cutting and sewing operations. Intelligent robotics have replaced traditional manual labour, allowing machines to perform tasks more accurately and quickly. This has not only improved product consistency but also significantly reduced labour costs. Automated systems can work continuously, increasing output rates and ensuring higher production volumes without sacrificing quality. Moreover, automation has substantially improved manufacturers' inventory management. Real-time tracking systems allow better control of stock levels and ensure that materials are utilized efficiently and with minimal excess inventory. This results in waste reduction and ensures production is closely aligned with the actual demand from consumers. With further advancements in automation, the efficiency of supply chains will increase even further, as technologies such as predictive analytics and AI will allow manufacturers to optimize production schedules, reduce lead times, and respond more quickly to changes in market conditions. Overall, automation is transforming the process of textile manufacturing to become more streamlined, cost-effective, and sustainable [22].

Despite technological advancement, challenges remain, including the high initial investment for machinery and continuous training to keep abreast of rapid changes. In the future, AI and machine learning will dominate trend forecasting, allowing designers to predict consumer preferences more accurately. Sustainability will be a focus area, driving innovations to reduce environmental impact. Technology is revolutionizing textile design through innovation, efficiency, and functionality while relevant to sustainability concerns. Adopting these changes determines success in the changing nature of the textile industry.

5 Conclusion

Exploring modern textile design shows a rich tapestry of evolution, innovation, and cultural significance. As already discussed, modern textile design is more than aesthetics—a mix of creativity, technology, and sustainability. It has shifted from the traditional to the contemporary; it reflects how production techniques have changed over time and increased consciousness of environmental and ethical concerns. Today, designers face much pressure to balance functionality with artistic expression and incorporate sustainable practices in tune with the consumers' values. The historical context also underscores the fact that the history of textiles has been that of storytelling and expression of culture. Textiles have reflected society's values, technology, and artistic movements from the earliest civilizations to modern times.

An important step has been taken by integrating these technologies for more customization and efficiency in production with digital technologies and intelligent textiles. Such innovations have helped designers create intricate patterns and functional fabrics catering to modern needs.

Moreover, the role of textile designers has dramatically changed in terms of work across disciplines and commitment to sustainability. Saving cultural heritage is paving the way for a more responsible future with fashion and interior design. Still, the industry faces some of the challenges related to waste and environmental impact. In conclusion, modern textile design is a potent tool for self-expression and cultural dialogue, balancing tradition with innovation. Honoured in their roots but stretched ever further, designers can imagine that the future is bright and diversified in the world of textile design. Values of sustainability and social responsibility are woven very intimately. This will constantly be in flux, changing the story of textiles in contemporary society and ensuring its place in this dynamic world.

References

1. Kiron, M. I. (2022, January 15). *Current trends in textile design*. Textile Learner. https://textilelearner.net/current-trends-in-textile-design/
2. Debeli, D., & Jiu, Z. (2013). Analyzing the cultural background of textile designers on their innovative thinking. In *International Conference on Education Technology and Management Science (ICETMS 2013)*. Atlantis Press.
3. Aiman. (2024, January 10). *Exploring the role of cultural textiles in our society*. Textile Mentor. https://textilementor.com/role-of-cultural-textiles-in-our-society/
4. Locofast. (2024, May 9). *The globalization of textile design: How cultures influence patterns*. https://locofast.com/blog/?p=1242
5. Kumar, M. (2023, February 2). *History of textile design*. Tutorials Point. https://www.tutorialspoint.com/history-of-textile-design
6. Wisniowski, M.-T. (2014, January 10). *Timelines of fabrics, dyes and other stuff*. Art Quill Studio. https://artquill.blogspot.com/2014/01/timelines-of-fabrics-dyes-and-other.html
7. Mayfairsilk. (2022, July 1). *The history of silk: The story of the luxurious fabric*. https://mayfairsilk.com/blogs/general/the-history-of-silk-the-story-of-the-luxurious-fabric?srsltid=AfmBOooG0os2-IPEcRHI28IqYtgV6CKif0NZyf5FvJAIUEjG-0ytH8KK
8. Editors. (2023, June 6). *Silk road*. History. https://www.history.com/topics/ancient-middle-east/silk-road
9. *A timeline of textile history*. (2022, November 27). Textile Heritage Museum. https://textileheritagemuseum.org/textiles-ancient-times-to-modern-day/
10. *File:Spinning jenny blackburn.JPG—Wikimedia commons*. (2023, May 1). https://commons.wikimedia.org/wiki/File:Spinning_jenny_blackburn.JPG
11. Un-Hjdm. (2024, September 30). *The evolution of textile design: From ancient techniques to modern innovations*. JD Institute of Fashion Technology. https://www.jdinstitute.edu.in/the-evolution-of-textile-design-from-ancient-techniques-to-modern-innovations/
12. The Art Story. (n.d.). *Textile art movement overview*. https://www.theartstory.org/movement/textile-art/
13. Soliwal, T. (2024, October 16). *Advancements in pattern engineering software in apparel industry*. Fibre2Fashion. https://www.fibre2fashion.com/industry-article/10161/advancements-in-pattern-engineering-software-in-apparel-industry

14. Audaces. (2023, December 7). *Understand the role of a textile designer in the fashion industry.* https://audaces.com/en/blog/textile-designer
15. Waddell, B. (2019, August 8). *The role of technology in the textile industry.* Textile Consult. https://www.textileconsult.co.uk/2019/08/08/the-role-of-technology-in-the-textile-industry/
16. Shah, D. (2023, April 18). *Impact of digital technologies on the textile industry.* Fibre2Fashion. https://www.fibre2fashion.com/industry-article/9622/impact-of-digital-technologies-on-the-textile-industry
17. Kamile. (2024, June 10). *14 Textile technology trends that are changing the industry.* Polygiene. https://polygiene.com/news/resources/textile-technology-trends/
18. *File:XYZprinting inBloom Dress 3D Printed Outfit (16837487086).jpg—Wikimedia commons.* (2015, January 7). https://commons.wikimedia.org/wiki/File:XYZprinting_inBloom_Dress_3D_Printed_Outfit_(16837487086).jpg
19. Younes, B. (2023). Smart E-textiles: A review of their aspects and applications. *Journal of Industrial Textiles, 53.* https://doi.org/10.1177/15280837231215493
20. Textile Insights. (2024, July 9). *The role of technology in the future growth of Indian textile industry.* Textile Insights—Delivering Inspiring Information. https://textileinsights.in/the-role-of-technology-in-the-future-growth-of-indian-textile-industry/
21. Fibreguard. (2023, September 19). *How technology is changing the textile industry.* https://fibreguard.com/sell/blog/how-technology-is-changing-the-textile-industry
22. Sood, S. (2024, October 25). *Revolutionising fashion: How automation is transforming the apparel industry.* Fibre2Fashion. https://www.fibre2fashion.com/industry-article/10182/revolutionising-fashion-how-automation-is-transforming-the-apparel-industry

Historical Context of Textile Design

Babita Bhandari

1 Introduction

Textile design is a vibrant and complex discipline to comprehend and analyze through timeline. Historically, it has expanded beyond geographical boundaries and evolved to varying extents through nomadic tribes, court ateliers and private workshops simultaneously [1]. Textile designs have been used as a means of expressing socio-cultural narratives, political agendas, religious themes, and human creative potential as well. Studies of textile history reveal that apart from textiles serving as protection from environmental conditions, these have been a great part of tribute to court, parts of rituals, trade culture, technology, etc. Moreover, textiles have been used to appreciate art and showcase strong belief systems, which is the appropriate mode of expressing allegations, preserving memories and proposing new ideas [2].

Textile design can be defined as the art and process of creating designs for woven, knitted, surface ornamented fabrics. It involves the creation of designs and patterns that are integral to the production of textile fabrics, influencing their texture, appearance, and properties [3, 4]. This process is deeply influenced by cultural, social, and economic factors, reflecting the heritage and identity of societies [5, 6]. Additionally, cultural elements and traditional motifs play a significant role in textile design that reflects the material and spiritual culture of a society [7, 8]. It is also influenced by cultural heritage, technological innovation, and market trends, all of which

B. Bhandari (✉)
National Institute of Fashion Technology, Kangra, Himachal Pradesh, India
e-mail: babita.bhandari@nift.ac.in

Sadhna et al. (eds.), *The Art and Craft of Modern Textile Design*, Sustainable Textiles: Production, Processing, Manufacturing & Chemistry,
https://doi.org/10.1007/978-3-031-86797-2_2

contribute to the evolution of textile patterns, structures, and the overall design process [9–11].

Since advances in textile design have occurred concurrently throughout many civilizations and societies worldwide, mapping textile design since the prehistoric age is a difficult endeavor. In the past, textiles were known not only for their utilitarian role but also for their decorative and many other usages. This chapter is a modest attempt to illustrate the historical background of textile design through important textile artifacts from key cultures. The exquisite artifacts are a true manifestation of the skills and techniques of artisans mastered and adapted over time. Numerous textile historians and academics have comprehensively documented the historical evidence of yarns, spindles, looms, and diverse textile-making techniques [1, 2]. However, as demonstrated by the ornate Byzantine silk, the European medieval tapestry, the Indian chintz and many more, this chapter elucidates the creative expression of artisans and the adaptability of textiles to assume the shape of any narrative.

2 Raw Materials Used in Early Textiles

The development of yarns, threads and cords of varying thicknesses and textures is considered to be the initial form of textile art. The earliest used materials as textiles were made from various fibres like raffia and leaf fibers apart from the skins and hides. Among all the plant sources, flax fibres were the first to be used for making different clothing materials, later followed by cotton and Jute etc. Other sources like reed, palm, rush, papyrus plant and flax were utilized to make fabrics, ropes, and other textiles. In the past, even without true spinning or without using actual looms, a vast range of textiles could be made like the variety of bark cloths made in Africa, South America and Indonesia [12, 13]. Figure 1 shows bark cloth dated the late nineteenth century from Southern Africa.

Fig. 1 Bark cloth handstitched with raffia thread, Southern Africa [13]

2.1 Wool

Sheep were domesticated in the uplands of northern Iraq as early as 9000 BC. The draft-spun woolen threads date only from 4000 to 3500 BC. Wool, already the principal fibre of Northern Europe by 700 BC, was enhanced by selective breeding [2, 14].

The finest of oriental wool cashmere was found in Syria at the latest by AD 200. Dyed wools were primarily reserved for cross-loop knitting, sprang, and the wefts in inlaid tapestry displaying both Greco-Roman and from the first century AD Christian Coptic imagery [2]. Wool has been primarily used in the making of carpets in Persia and Turkey, as well as garments and other textiles in Europe and Central Asia.

2.2 Silk

The earliest found silk fragments are dated to about 3036 BC and are composed of *Bombyx mori* filaments originating from East-central China (Henan Province). The earliest surviving silk cloths made from the filaments produced by *Bombyx mori* are fragments dated to about 3630 BC and come from Henan province in east-central China. The artifacts found during the Neolithic era (5000–1700 BC) also suggest the widespread use of cocoons and silk. Silk was found amid excavated royal burial grounds associated with the Shang dynasty and the Western Zhou. Silk was interwoven with small geometric patterns and dyed in rich colors at least as early as about 1100 BC [15]. Silk carpets from Persia, silk textiles from Europe and silk tapestries from China have been a testament to artisanal beauty and exquisiteness. Until the fifth century AD, silk was being imported to major parts of the world from China due to unawareness of silk-rearing practices.

2.3 Linen

Linen fabric dated 2400 BC with indigo-dyed borders was excavated from Egyptian tombs. It is claimed that it was a crucial element of everyday life in Ancient Egypt and was used for clothing, bedding, and sails. It was cultivated from flax, a plant native to the region [16]. As per documentary evidence, linen was widely used fibre in Egyptian cultures as beddings and hangings and a heap of it was required for mummification with a great variety of thicknesses and textures of plain weave predominantly. The finest cloth was made from the earliest times. The world's oldest woven garment displayed in Fig. 2 has been confirmed to be made of Linen dated between 3482 and 3102 BCE, and it is believed to be over 5000 years old [17]. The Pharaonic era in Egypt also emphasized the importance of linen, with weavers

Fig. 2 The Tarkhan Dress [17]

addressing the scarcity of linen by utilizing other natural materials, highlighting the sustainability practices of the time [18].

2.4 Cotton

Cotton seeds and textiles were found in the Arabian Peninsula at *Qal'at al-Bahrain and Mada'in Sâlih* as a center of textile production, suggesting cotton was grown for trade and possibly competed with other Middle East sources [19]. Spun yarns of mixed cotton and wild silk were found in India around 1500–100 BC [2].

Indian cotton was also renowned for its designs—brilliantly colored and patterned, dyed and printed to the tastes of innumerable communities. These clothes were known and valued across much of the world from ancient times, from the South China Sea to the Mediterranean [20].

2.5 Natural Dyes

Natural dyes in historic textiles include yellow, brown, blue, and red dyes, and mixtures of two dyes were used to obtain desired shades [21]. Early evidence shows that textiles were dyed with plant-based dyestuffs like *Rubia tinctorum L.* and indigo in the Late second millennium BCE [22]. The indigo blue found at the Preceramic site of Peru suggests the first use of indigo dye, as seen in 6000-year-old cotton fabrics

[23]. Ancient Egyptian textiles from the Roman and Byzantine periods show the application of mollusc purple, indigo or woad, madder, and as well as kermes, cochineal, and lac [24].

Evidence of the use of locally grown madder for red shades and imported indigo for blue shades is witnessed at the Yanghai archaeological site in Late Bronze Age textiles. These represent the oldest record of madder and indigo textile dyeing in the Xinjiang Region [25].

Other valuable dyes used in ancient times were red dyes derived from some scale insect species. Sources of yellow dyes were weld, saffron or turmeric [1]. Indigo, a source of blue shades, was cultivated and exported in India. Woad, more common in Europe and the Near East, was also dyed for blue dye. A great collection of indigo-dyed fabric fragments is kept in the Victoria and Albert Museum, London, dating from the Chinese Han to Tang dynasties. In ancient Israel and Palestine, indigo was combined with green and black dyes. In the Middle Ages, the textile manufacturers of Europe used Woad, locally grown in abundance, to supply the flourishing weaving industry. Until 1498, it was the main source of blue dye. With the increase of imports, mainly from India, woad died out as a major crop in the mid-eighteenth century but continued on a small scale [26].

3 Tools and Techniques Used in the Earliest Textiles

Initially, only tablets, disc or simple frames were reported to be used for weaving earliest textiles. Warp-weighted looms were used, as evidenced by the loom weights found during the period 526–332 BCE. Before AD 250, the first horizontal loom was developed in the Eastern Province by Roman weavers and beautiful geometric patterns in damask silks were woven with horizontal looms [27].

Different types of looms were used for weaving in the Near East and Eastern Mediterranean [28].

The upright frame loom was invented in the first century AD to facilitate weft wrapping techniques and tapestry. Before the Han Dynasty, plain weave and figure-weave fabrics were produced on similar looms, while after the Han Dynasty, looms evolved into various types, including treadle looms, multi-heald and multi-treadle looms, and draw looms [29].

The much simpler Greek attire is often depicted in sculptures and vases of the sixth and fifth centuries BC, as occasionally the warp-weighted looms show decoration that is normally limited to starting borders. Pit looms and backstrap looms have been popularly used in Asian cultures.

4 Design Evolution in Woven Textiles

Indigenous genius fuelled by a complex belief system, coupled with widespread and historic trading contacts and a history of migration, has given rise to a wide variety of patterning techniques and myriad design motifs in textiles around the world.

The history of textiles, from prehistory to the early twenty-first century, demonstrates the importance of human ingenuity and global trade, shaping the modern textile design [2].

The earliest design of tapestry techniques dates back to around 1400–1330 BC in Egypt in which the surrounding cloth was plain woven linen, but the inserted imagery was coloured wool [2].

Tapestry fragments suggest that those were used as dress ornaments in Egypt, as depicted in Fig. 3. Geometric patterns in mythological hunting scenes often predominate with wine scrolls. The weft loop pile technique was also used for pictorial textiles.

Figure woven and plain Chinese silks were found in the burial mounds of the Altai Mountains of Kazakhstan, at the Scythian site of Pazyryk, dating back to 500–400 BC. Various types of materials and techniques were used or present in the region of Gordium during the eighth century BC as suggested by the findings of various decorative and utilitarian textiles such as soumak, coloured stripes, checks, slit tapestry, weft loops, compound weft-faced patterns using off-white vegetable fibers and purplish red wool. Alongside plain-woven cloths of flax and wool, there were fabrics of mohair and hemp as well.

Among the textiles preserved in the frozen tombs of southern Siberia were pile carpets and fragments dating from the sixth and fifth centuries BC [1].

Fig. 3 Fragment, ornament from a tunic, tapestry woven linen and wool, Akhmim, Egypt, fourth century to sixth century [2]

In China, various Tomb excavations have found plain weaves, gauze, embroidered silk, leno, and qi-damask. The elaborate woven patterns depicting intermingling images of dancing figures, dragons and other beasts among geometric motifs suggest that an advanced loom was used. With patterning rods and shafts operated by several people, the treadled pattern-rod loom, later developed into known as pattern-frame loom or Chinese draw loom, provided the mechanism for the even more intricate weaves for which China was soon to become renowned [2].

Several variations of patterns with Chinese characters dated as early as 150 BC have been found along the silk routes in Xinjiang. The spindle wheel developed for the bast fibers was adapted for silk during the Zhou period. The earliest known silk threads are incorporated as stripes or supplementary wefts in woven cloths or used as fine embroidery, all in local styles and accompanied by other complex fabrics such as elaborate tablet woven borders and complex twills. Trading in silk reached as far as Rome and beyond to Africa through various silk routes linking the regions in commerce in the ancient world till the fifteenth century. Decorative worsted twills, gold brocaded tablet woven bands, and sumptuous silk gold embroideries are found in AD 410–76.

In both Egypt and Persia, then the two dominant centres for weaving sophisticated textiles, a system of state workshops was long established. The weaving workshops that produced luxurious textiles not only for the court but also for the wealthy people were termed *Tiraz*.

Historically, silk was the highest-valued textile commodity in the world. From early times, it had become a crucial item of trade for both the Byzantine and Sassanian empires. Techniques of weaving mulberry silk had been smuggled out of China, reaching Byzantium by the sixth century AD. The Caspian coast of Persia became the highest-producing raw silk in the non-Chinese world. Eventually, figured silk in complex techniques such as lampas and samite was produced by the weavers of Persia, Egypt, Baghdad and Moorish Spain. Silk textiles have been the most influential, as they were a key trade commodity and important diplomatic gifts exchanged between empires [14].

Sassanid textiles (AD 224–651) subsequently influenced the design of textiles from Spain to Japan. These motifs were also found in China, Central Asia, France and the Byzantine Empire. Hunting was an obsession of the Persians and was closely linked with their religious beliefs. Their iconography of kinship, battles and hunting and their heraldic imagery, appealed outside Iran to feudal societies as widespread and varying as those in early Islam, in Christian Europe and in Tang China. Woven designs were based on animals like ibex, lions, cockerel and other birds, as well as fabulous beats and senmurv (a dog with a tail of peacock) as shown in Fig. 4, the griffin (part eagle, part lion) and the Pegasus (a winged horse). These animals were adorned with fluttering ribbons, jewelled borders, adorned with pearled medallions. During Byzantium era, silk was woven in tabby, twill, lampas, and tapestry weave. These textiles are known for their bright colours, intricate designs and use of gold threads [1]. By the twelfth century, Nakshi, a more cursive script, was becoming popular and was incorporated into complex silks. At the end of the twelfth century, roundels were superseded by ovals and hexagonal frames,

Fig. 4 Fragment of a silk with a woven-in pattern showing a large composite beast in a roundel with a pearl border, Iran or Central Asia, seventh or eighth century [1]

and set within these were animals and birds, both real and mythical in either addorsed or confronted pairs. By the thirteenth century, the figurative motifs of humans and animals inherited from previous civilizations were replaced by arabesques, geometric patterns and floral motifs in Islamic textiles.

The heraldic imagery of hunting scenes and beasts both real and imaginary decorating these Islamic export textiles appealed to the prevailing European aesthetic, which seems to have differed little from that of the Muslims. Patterns during the thirteenth century in Spanish silks include stars, rosettes and square grids composed of interlaced vertical, horizontal and diagonal axes.

In the fifteenth century, weavers preferred pomegranates, palmettes, and lotuses as textile motifs often used in combination with birds and animals.

In the mid-fifteenth century, the first weaving center in France devoted to high-quality silks, especially velvets, was initiated. The predominant textile motif of the Italian Renaissance was the pomegranate, as shown in Fig. 5, which was introduced into the design lexicon through trade with the Ottoman Empire. Closely associated with the noble families and their courts, the pomegranate symbolized eternity, fertility, and resurrection.

Fifteenth- and early sixteenth-century silks from Granada patterned in a series of bars reminiscent of certain tiled mosaics from the Alhambra palace known as Alhambra silks (Fig. 6). A common design was rows of stars made up of interlacing. By the sixteenth century, the principle figured weaves-velvets, lampas, and damask with brocaded effects had been developed. Diverse designs in terms of scale, types, etc., were being produced in different weights of textiles. The ogee lattice is the principal design composition of the sixteenth century.

In Safavid Iran (1499–1722), lampas silk, in many colours and metal threads, were drawn with such virtuosity that they competed with penwork. Narrative scenes were also woven in velvet with fine-cut piles and voided satin grounds. Court artists, architects and craftsmen were expected to be versatile and painting and book design, textile manufacture and carpet weaving flourished side by side. Round sixteenth-century Ottoman weavers produced, for court use, silk and precious metal brocades and velvets often voided with distinctive floral patterns influenced by newly

Fig. 5 Cut and voided velvet panel with metallic threads, Italy, fifteenth century

Fig. 6 Silk lampas, South-East Spain, 1300–1400

fashionable Italian silks. Safavid Persia under Shah Abbas became a world-renowned supplier of figured silk and velvets (Fig. 7).

The velvets with images of beautifully drawn hunting scenes, picnics and drinking parties set against a golden yellow were world famous and highly valued when received as diplomatic gifts, whether in Europe, Russia, Japan or Siam, brought into India from Europe favoured flowers standing alone, or in rows or three.

Mughals were also famed for their delicate muslins, the products of Bengali looms around Dacca and in parts of northern India around Lucknow and Benaras. The finest Brocades known as kinkhabs were also woven in Gujrat. During the

Fig. 7 Silk panel, twill figure on satin ground, Iran, 1550–1600

seventeenth century, from about 1630, woven flowers became much more realistic and stylized curvy leaves and stems inspired by peony and chrysanthemum. By the mid-seventeenth century damask had created a dynamic play of light on the background surface. A different weave was also developed, i.e., a compound plain weave with inner warps and extra wefts. A new interesting colour palette, i.e., lime green, turquoise blue, salmon pink, lemon yellow, etc., was derived by mixing different dyes. Persian paintings from the 1736–1834 period show the presence of carpets, felts, and silks as floor coverings, impacting carpet weaving in the post-Safavid period [30]. Central Asian carpet weaving history reveals the development of materials and decorations, as well as the history of the types and names of rugs [31].

A common visual language was established in textile design by the mid-seventeenth century owing to the exchange of textile motifs, patterns, fibres and dyes in the wide textile trade. It was being recognized throughout the world. Tapestries in the Renaissance were a key element in the ostentatious magnificence used by powerful rulers to broadcast their wealth and might, contributing to art, liturgy, and propaganda during the period. One example of such tapestries is given in Fig. 8 [32].

5 Design Evolution in Embroidered Textiles

The earliest examples of embroidery (1800 BC) come from Anatolia, where pure gold wire was used to secure dark blue beads. The earliest surviving embroidery is Egyptian around 1400–1300 BC. Embroidery also has a long history in China, dating from the fourth century BC onwards and European and Islamic embroidery around 4000–2000 BC [2]. The Bayeux Tapestry is a unique artefact from the

Fig. 8 Tapestry woven in wool and silk on wool warp, 1475–1490 [32]

eleventh century, revealing the political and social history of the Battle of Hastings and its makers [33].

Indian embroideries made for the trade were considered to be luxury textiles. Their designs were customized to suit the taste of the target foreign market. Indian trade embroidered for Britain merged the elements from Indian and European aesthetics and were imported from India between the seventeenth and nineteenth centuries for use as dress materials and furnishing materials. Embroidery in Europe has evolved from domestic art to commercial one, with England being the most important domestic embroidery producer in the seventeenth century and France producing sophisticated embroidery since the sixteenth century. Embroidery became more finely detailed in the eighteenth century, with France being the leader in using finer wool and silk threads for domestic and costume embroidery [34]. Embroidery and needlework history includes medieval English vestments, Baroque, Rococo, neoclassical, and modernist styles, influenced by Chinese embroidery (Fig. 9) and Indian textiles [35].

Indian textile designs have evolved overtime, primarily influenced by nature, architecture, culture, and religion, and have become an intricate art form [36].

6 Lace

In the late fifteenth and early sixteenth century, bobbin and needle lace were made. Metal threads and silk were used as special materials for lace making in Italy and Venice. Different types of trimmings were produced working with bobbins like many types of plaits and braids. The spread of pattern books had an important influence on the development of lace in sixteenth century. Printed pattern books

Fig. 9 Panel, embroidered silk, probably Guangzhou (Canton), China, eighteenth to nineteenth century

including pulled thread, drawn thread, cutwork and lace patterns were being produced during the next century [37].

7 Design Evolution in Dyed and Printed Textiles

Huge leap of textile fragments of Indian origin unearthed in Fustat region of Egypt suggest the highly developed craftmanship of Indian dyers and the design aesthetic suitable for everyday use. Chintz and Indian printed palampore (Fig. 10) describe the pinnacle of hand-dyed and painted techniques expertised by Indian craftsmen, which were in rage in eighteenth century England. Drawings of specific patterns and pattern sheets were being sent to Indian artisans and those were being customized for European consumers as per their desired colour palette and design aesthetics [38].

The patterned clothes were dominated by combination of red, blue, black, violet, green, yellow. Indigo was the principal dye together with chay and madder red in the decoration of vast quantity of Indian cotton goods from at least the late fifteenth century. Its popularity surged in the seventeenth and eighteenth centuries when European traders introduced it to the West, where it became highly coveted for its exotic beauty and luxurious feel [39]. Through these printed patterns incorporating flora, fauna, geometric and heraldic forms multifaceted discourse enriched the visual and material resources of much of the world.

Printing was practically not known in the West before seventeenth century. In the second half of eighteenth century printing from copper plates were popular. It

Fig. 10 Palampore of painted and dyed cotton chintz, Coromandel Coast, ca. 1720–1740

Fig. 11 Printed cotton, about 1785, French; Plate printed. Oberkampf at Jouy-en-Josas; Grape Harvests; red on white

creates much larger designs with intricate details. From 1760 to 1800 some of the finest printed textiles were made using this technique popularly known as toile de Jouy (Fig. 11). Pattern dyeing in the form of wax or clay resist fabrics have been found in Egyptian burial grounds of fifth to sixth centuries AD and resist-dyed silks have been found in probably imported from China (AD 646–749). Another resist dyeing is tie-dyeing known as bandhana in India, shibori in Japan, plangi in Indonesia, adire in South America, Central Asia, and Africa. Yarn tie dyeing known as Ikat are known as Kasuri in Japan, Geringseing in Indonesia, Bandha, Patola in India [1].

8 Inspirations in Historic Textile Design

The inspirations for traditional textile designs have different sources based on the cultural integration of various elements by the community. These are sometime governed by the underlying meanings in the sources of designs and are reinterpreted in various forms in terms of patterns, colours and decorations etc. Motifs are sometimes credited with powers of sympathetic magic. Hunting cloths made around Herat are stenciled, printed and block-printed with red and mythical beasts. Scorpions embroidered on a pair of long woolen puttees are believed to protect the legs of the wearer from live insect in Central Asia. Ram's horn was considered ancient emblem of success and strength in hunt [40]. Variety of primitive symbols were used as reference for design elements like pomegranates, fish, Anthropomorphic figures, spirals, carnations and tulips, plants, 8-pointed stars and birds, squares, triangles, circles, swastikas, whorls, hooks zigzag, crosses, waves, dots etc. Moreover, religious concepts were displayed in the form of certain insignias [41]. Table 1 illustrates the various categories of inspirations for derivation of different motifs and patterns found in historic textiles across cultures.

9 Ancient Textile Traditions Across Cultures

Tradition does not remain the same. It evolves which time. Various factors wich influence the tradition are community, hardship or prosperity, networking with the outside community and climate change etc. Some traditions are universal, which are

Table 1 Inspirations for ancient motifs in different cultures [1, 2, 12, 14, 36, 42]

Inspirations	Motifs/designs
Mythical	Senmurv (a dog with a tail of peacock), the griffin (part eagle, part lion) and the Pegasus (a winged horse), Dragon, Pheonix, unicorn, royal ram
Religious	Coptic, Biblical, Islamic, Buddhist, swastikas or auspicious symbols
Narrative/ thematic	Hunting scenes, the lady and unicorn, courtly scenes, historical event. countryside landscape, romantic scenes
Geometric	Medallion, square, triangle, oval, circle, hexagon
Trees	Palm, tree of life, bamboo, pine, plum tree
Floral and foliage	Tulip, chrysanthemum, iris, tulip, carnation, daisy, lily, rose, Palmette, lotus, acanthus
Birds	Peacocks, cranes, sacred geese,
Animals	Geese, lion, tiger, deer, camel, fish, elephant,
Fruits	Pomegranate, grapes, cherry, plum
Figurative	Huntsmen, warrior, dancers, king, queen
Architectural	Pagoda, cartouche, eight-pointed star, arabesque designs, arched, scalloped, columns
Tribal	Tribal fingers, adinkra symbols, tribal motifs

observed in most of the cultures. The use of expensive materials like silk and metal threads or elaborate weaving and embroidery outfits were designed to display one's wealthy status. All over the world, to protect the wearer from evil eye or as good luck, some magical designs are embroidered on the textiles. Traditions and textiles evolve, as the community evolves. Various indigenous communities have developed and maintained their complex and distinctive textile traditions [43].

Textiles were utilized as currency in the past owing to their rich raw materials, consisting of silk, fine cotton, pashm-wool, and colors like cochineal, lac, and Tyrian purple. Textiles were prized for their ability to purchase or persuade, as well as their prestige, aesthetic appeal, and physical strength.

In some indigenous communities, textile patterns convey ancestral knowledge and spiritual connections to the natural world, emphasizing the deep-rooted cultural significance of weaving and embroidery [42]. Some significant textile traditions are mentioned below:

- Embroidery was used to serve to identify the status of the people. It was used to indicate the village to which she belongs, and the colours and position of the design indicate the status of women as young, unmarried,/widowed women.
- Moreover, special occasions are celebrated with embroidery. At all rites of passage when people are considered most vulnerable to spirits, embroidery plays a symbolic role. In ancient societies death was the most important, more recently marriage.
- In Romania embroideries are hung outside the house of marriageable girl.
- In Turkey the fineness of embroidery on the towels a young girl takes to the public baths announces her quality as a prospective daughter-in-law.
- At childbirth the passage of 40 days is marked with protective embroidery for mother and baby, and death. The embroideries worn at marriage serve as a shroud.
- The ceremonial sarongs of southern Sumatra are worn at all rites of passage and often among the textiles buried with the dead.
- Headhunting was a deeply symbolic practice to the Naga people. To go on headhunting expeditions, the men wore Killing shawls decorated with moon and anthropomorphic motifs in cowrie shells with squares of red dog's hair.
- In North Africa, after consummating the marriage usually on the seventh day, the ceremony of putting on the belt symbolizes the bride's return to normal life. She wears an enormous woolen shawl into which the symbolic fish is also woven, half henna dyed and half indigo with an embroidered center panel. This is tied around with a belt in the presence of small boys to ensure she will have sons.
- In Chinese courts, officials used embroidery badges on their coats to signify their ranks in different and animal motifs.
- In Japanese textiles, the hemp leaves play a talismanic tool in the clothing of babies and children.
- In Ghana, Africa, Talismanic charms sewn on to the clothing were used by hunters warriors and also shamans. The shamans originating in the hunting societies of the palaeolithic era, also identify himself with the animal world by decoration, for the hunter the amulets afford protection.

- In India, several head shawls are skillfully embellished to make up her bridal trousseau. In Gujrat, bride is draped in a special red sari having checkered pattern with tie and dye motifs known as gharchola during the wedding ceremony.
- The Patola is believed to protect the mother and ensure a safe birth.
- Decorated textile is used as canopy to cover the bride and groom during marriage rituals near Samarkand.
- Silk owing to its strength and ability to take dyes readily was preferred choice by the warriors and was used in tents, protective garments, flags etc.
- At Mecca, the shrine of Muhammad, the Ka'ba is covered by a gold black silk cloth with passages from the Koran. It is renewed annually and pieces of the old one are distributed to pilgrims.
- Cloth layout and motif arrangement might denote family lineage and clan identity.
- Special sacred cloths were woven according to a set of strict taboos for religious ceremonies.
- Certain weavings conferred invisibility on warriors during battle, while others protected against illness and misfortune in South East Asia.
- In Turkey, beautiful cloths with either embroidered or woven decoration were also used to wrap gifts or to store household items.
- Kesi tapestries were also used as diplomatic gifts and tributes, exchanged between rulers as tokens of esteem and goodwill.

10 Impact of the Industrial Revolution on Textile Design

The First Industrial Revolution profoundly impacted textile design, majorly through the mechanization of production processes. The introduction of steam power and the industrialization of textile production, marked the beginning of significant changes in how textiles were manufactured. This period saw the transformation of the textile industry from a manual, labour-intensive craft to a mechanized, efficient production system [44, 45]. The most significant developments in weave production started in the eighteenth century when the Industrial Revolution pioneered a shift towards mechanical production methods that undermined the manual craft of weavers in Europe. Important textile tools like spinning jenny, water-powered spinning machine, water frame, spinning mule were invented by James Hargreaves, Richard Arkwright and Samuel Compton respectively enabling effective spinning of yarns.

During nineteenth century advancement in chemistry led to the evolution of textile design in terms of dyeing and printing. Important dyes like mauvine and indigo dyes were synthesized out of aniline and quinine respectively. More regenerated man made fibres like viscose was invented by Count Hilaire de Chardonnet in 1884 and a factory was dedicated to the production of viscose that acted as a replacement of natural silk in a cost effective manner.

Interestingly, while the First Industrial Revolution facilitated mass production and consumption patterns, it also influenced the design aspect of textiles. The scale of investment and centralization of production, as exemplified by Benjamin Gott's operations, led to technological and organizational innovations that propelled the mechanization of the wool textile trade [46]. This shift allowed for the creation of new textile designs and patterns that could be produced on a larger scale, thus affecting the processes used and steps in textile design [4].

11 Conclusion

Textile fragments found in different archaeological sites and heritage textiles conserved in textile museums provide a great idea of the vastness and richness of textile crafts practiced around the world. The vibrant craftsmanship of the wide range of heritage textiles has always inspired designers and artists alike and has continued to do so for millennia. Historic textiles laid the foundation for the artistic aesthetic aspect of textiles, including a thorough understanding of the materials, processes, designs, and pattern layout customized for different usages. Evolution and developments in textile designs over the period in different cultures owing to the introduction of various fibers, tools, techniques, and craftsmanship serve as great sources of inspiration for the industry, academia, and art connoisseurs alike. The prevalent traditions associated with the traditional usage of textiles in different communities represent a strong cultural identity for future generations which further reinforces the cultural integrity. The historical context of textile design will play an instrumental role in establishing a relationship to the cultural roots and further imbibing sustainability ethos into the textile processes as the textile design field continues to evolve.

Acknowledgements The author would like to acknowledge the support provided by the Resource Center, NIFT Kangra, Himachal Pradesh during the study. The credit for image sources is given to the Victoria and Albert Museum, London's textile collection.

References

1. Harris, J. (Ed.). (1993). *5000 Years of textiles*. British Museum Press.
2. Mitchell, V., & Schoeser, M. (2003). *World textiles: A concise history*. Thames and Hudson.
3. Wilson, J. (2001). The principles and elements of textile design. In *Handbook of textile design*. Woodhead Publishing.
4. Ujiie, H. (2002). Textile design education in digital inkjet fabric printing. In *Proceeding IS&T International Conference on Digital Printing Technologies (NIP18)* (pp. 254–257). Springer.
5. Kraft, K. (2004). Textile patterns and their epistemological functions. *Textile, 2*(3), 274–289. https://doi.org/10.2752/147597504778052711
6. Studd, R. (2002). The textile design process. *The Design Journal, 5*(1), 35–49. https://doi.org/10.2752/146069202790718567

7. Abdullah, Z., Musa Kahn, S., Yahaya, S. R., Harun, A., & Ghazali, R. (2018). The potential of floral pattern art in Malay textile embroidery: A formalistic analysis. *Jurnal Pengajian Melayu (JOMAS), 29*(1), 286–305.
8. Miao, H. Q., Nie, K. W., & Ding, Z. P. (2013). Discussion on application of the Chinese favonian patterns in silk home-textile design. *Advanced Materials Research, 796*, 255–258. https://doi.org/10.4028/www.scientific.net/amr.796.255
9. Chatterjee, A. (2021). Life, after life: Textile crafts in India and communities of practice. *Textile, 20*(3), 245–251. https://doi.org/10.1080/14759756.2021.1993558
10. Brown, S., & Vacca, F. (2022). Cultural sustainability in fashion: Reflections on craft and sustainable development models. *Sustainability: Science, Practice and Policy, 18*(1), 590–600. https://doi.org/10.1080/15487733.2022.2100102
11. Dubois Athenor, P. H., Hintzen, N., Igarashi, N., Ramchandani, M., & Coste-Manière, I. (2022). Traditional textiles going local and global. In S. S. Muthu (Ed.), *Sustainable approaches in textiles and fashion. Sustainable textiles: Production, processing, manufacturing & chemistry*. Springer. https://doi.org/10.1007/978-981-19-0874-3_7
12. Klarke, D. (2002). *The art of African textiles*. Thunder Bay Press.
13. *African barkcloth.* https://www.nms.ac.uk/collections-research/collections-departments/global-arts-cultures-and-design/projects/african-barkcloth/
14. Gillow, J. (2010). *Textiles of the Islamic world*. Thames and Hudson.
15. Schoeser, M. (2007). *Silk*. Yale University Press.
16. Elsharnouby, R. (2014). Linen in Ancient Egypt. *EKB Journal Management System, 15*, 1–22. https://doi.org/10.21608/JGUAA.2014.3087
17. Lobell, J. A. (2016). *Dressing for the ages. Archaeology*. Archaeological Institute of America.
18. Rehab, E. (2020). *Linen in ancient Egypt*. https://www.researchgate.net/publication/340595105_Linen_in_Ancient_Egypt#fullTextFileContent
19. Bouchaud, C., Tengberg, M., & Prà, P. (2011). Cotton cultivation and textile production in the Arabian peninsula during antiquity; the evidence from Madâ'in Sâlih (Saudi Arabia) and Qal'at al-Bahrain (Bahrain). *Vegetation History and Archaeobotany, 20*, 405–417. https://doi.org/10.1007/S00334-011-0296-0
20. Lemire, B. (2011). *Cotton*. Berg.
21. Kharbade, B., & Agrawal, O. (1988). Analysis of natural dyes in Indian historic textiles. *Studies in Conservation, 33*, 1–8. https://doi.org/10.2307/1506235
22. Sukenik, N., Iluz, D., Amar, Z., Varvak, A., Workman, V., Shamir, O., & Ben-Yosef, E. (2017). Early evidence (late 2nd millennium BCE) of plant-based dyeing of textiles from Timna, Israel. *PLoS One, 12*, e0179014. https://doi.org/10.1371/journal.pone.0179014
23. Splitstoser, J., Dillehay, T., Wouters, J., & Claro, A. (2016). Early pre-Hispanic use of indigo blue in Peru. *Science Advances, 2*(9), e1501623. https://doi.org/10.1126/sciadv.1501623
24. Karapanagiotis, I., Verhecken-Lammens, C., & Kamaterou, P. (2018). Identification of dyes in Egyptian textiles of the first-millennium ad from the collection fill-Trevisiol. *Archaeological and Anthropological Sciences, 11*, 2699–2710. https://doi.org/10.1007/s12520-018-0705-8
25. Kramell, A., Li, X., Csuk, R., Wagner, M., Goslar, T., Tarasov, P., Kreusel, N., Kluge, R., & Wunderlich, C. (2014). Dyes of late bronze age textile clothes and accessories from the Yanghai archaeological site, Turfan, China: Determination of the fibers, color analysis and dating. *Quaternary International, 348*, 214–223. https://doi.org/10.1016/J.QUAINT.2014.05.012
26. Prideaux, V. (2003). *Book on Indigo: A handbook of Indigo dyeing*. Search Press.
27. Wild, J. (1987). The Roman horizontal loom. *American Journal of Archaeology, 91*, 459–471. https://doi.org/10.2307/505366
28. Ellis, R. (1976). Mesopotamiam crafts in modern and ancient times: Ancient near eastern weaving. *American Journal of Archaeology, 80*, 76–77. https://doi.org/10.2307/502941
29. Li, B., Liu, A., Li, Q., & Yang, X. (2012). Study on the evolution of the looms in ancient China. *Advanced Materials Research, 627*, 449–455. https://doi.org/10.4028/www.scientific.net/AMR.627.449

30. Maktabi, H. (2009). Under the peacock throne: Carpets, felts, and silks in Persian painting, 1736–1834. In *Muqarnas* (pp. 317–348). Brill. https://doi.org/10.1163/ej.9789004175891.i-386.81
31. Kuryazov, U. (2019). From the history of the development of carpet weaving. *Journal of Look into the Past, 21*(2), 38–42. https://doi.org/10.26739/2181-9599-2019-21-05
32. Campbell, T. (2002). *Tapestry in the renaissance: Art and magnificence*. https://doi.org/10.5860/choice.40-1313
33. Musset, L., & Rex, R. (2011). *The Bayeux tapestry*. https://doi.org/10.1093/nq/s1-xi.283.245f
34. Lee, K. (2009). Transition of embroidery in Europe. *Fashion & Textile Research Journal, 11*, 231–241.
35. Synge, L. (2001). *Art of embroidery: History of style and technique*. Woodbridge. https://doi.org/10.5860/choice.39-2577
36. Mehta, R. (2020). Evolution of Indian of Indian textile design: A study over timespan. *IARS International Research Journal, 10*(2). https://doi.org/10.51611/IARS.IRJ.V10I2.2020.146
37. Browne, C. (2004). *Lace from Victoria and Albert Museum*. V A Publications.
38. Thompson, A. (2007). *Textiles of South East Asia*. Crowood Press.
39. Percival, M. (1923). *The chintz book*. William Heinemann. https://archive.org/details/chintzbook0000maci/page/n3/mode/2up?view=theater
40. Harvey, J. (1996). *Traditional textiles of Central Asia*. Thames and Hudson.
41. Askari, N., & Arthur, L. (1999). *Uncut cloth*. Merrell Holberton Publishers.
42. Paine, S. (2010). *Embroidered textiles*. Thames and Hudson.
43. Williamson, L. (2016). Tracing textiles, motifs and patterns: Historical to contemporary. In *Textile Society of America Symposium Proceedings* (Vol. 1014). http://digitalcommons.unl.edu/tsaconf/1014
44. Watt, M. (2000). Nineteenth-century European textile production. In *Heilbrunn timeline of art history*. The Metropolitan Museum of Art. http://www.metmuseum.org/toah/hd/txtn/hd_txtn.htm
45. da Motta e Albuquerque, E. (2023). The initial impacts of the industrial revolution: An "astonishing reversal"—1771–1850. In *Technological revolutions and the periphery. Contributions to economics*. Springer. https://doi.org/10.1007/978-3-031-43436-5_3
46. Smail, J. (1999). Conclusion: Technology, innovation and the industrial revolution. In *Merchants, markets and manufacture*. Palgrave Macmillan. https://doi.org/10.1057/9780230513600_8
47. Shishlina, N., Orfinskaya, O., & Golikov, V. (2003). Bronze age textiles from the North Caucasus: New evidence of fourth millennium BC fibres and fabrics. *Oxford Journal of Archaeology, 22*, 331–344. https://doi.org/10.1046/J.1468-0092.2003.00191.X
48. Yarmolovich, V. (2022). Loom weights from archaeological excavations in Memphis: Some considerations on dating of vertical loom in Egypt. *Vostok. Afro-aziatskie obshchestva: istoriia i sovremennost*. https://doi.org/10.31857/s086919080020619-7
49. Roth, H. (2010). Studies in primitive looms. *Nature, 102*, 346–346. https://doi.org/10.1038/102346a0

Sustainable Practices in Textile Design

Babita Bhandari

1 Introduction

Before the industrial revolution, textile production was a labour and time-intensive process. The raw materials used in textile production were mainly natural and various fibres came only from nature, and laborious, time-consuming processes were used to process the raw materials into fabrics. People practiced sustainability without realizing it. With the mechanisation, fabric production started taking place in bulk. From the late nineteenth to mid-twentieth century, numerous types of manufactured fibres such as acrylic, nylon, acetate, polyester, etc., started flooding the marketplace [1].

The textile and apparel industries stand next to agriculture in contributing to water contamination with chemicals released from various processes used in textile finishing like scouring, bleaching, dyeing etc. In the 1990s, a New Jersey company, Wellman Inc., started recycling PET bottles for fibre production. Some of these fibres, labelled Fortrel®, Ecospun™, appeared in the new fleece fabrics. Other companies also started researching renewable raw materials such as corn fibre and soybeans [1]. However, with increasing access of textiles to various consumer classes, threats to the environment and unfair labour practices accelerated.

Sustainable development, as explained by The Brundtland report, caters to the needs of the people and other species on earth while taking care of future generations. There are many definitions of sustainability which talk about having a holistic approach and express concern about addressing the other issues which are also crucial for a sustainable future. As mentioned in the UN World Summit report, the

B. Bhandari (✉)
National Institute of Fashion Technology, Kangra, Himachal Pradesh, India
e-mail: babita.bhandari@nift.ac.in

Sadhna et al. (eds.), *The Art and Craft of Modern Textile Design*, Sustainable Textiles: Production, Processing, Manufacturing & Chemistry,
https://doi.org/10.1007/978-3-031-86797-2_3

action plan must consider the three pillars of sustainable development, social, economic, and environmental, to reach a solution [1].

2 Environmental Impact of Textile Fibers

Different materials, either natural or synthetic, impose various degrees and types of impact on ecology and society [2]. As per the Textile Exchange report, 2023, the production and extraction of raw materials contributes 24% to the total GHGs produced by the textile industry, which negatively impacts the communities and ecosystems [3]. Sustainability has become a long-discussed issue now.

The sustainability pioneer, Kate Fletcher, discusses various aspects of sustainability in detail including materials, processes, technology, workers, etc. [4].

All natural fibres are, by default, considered sustainable materials [5–7]. In recent years, renewable raw materials have been presented as sustainable materials by various brands; however, renewable raw materials do not ensure the sustainability of the final product. It relies on multiple factors such as conditions in which materials were grown, how much energy, water and other resources were utilized in different manufacturing stages, how it will impact the ecosystem and social system, and ultimately how it will be used by the consumers and discarded [2].

Careless production and extraction of natural fibres could also harm the environment; besides this, hazardous chemicals are used in various processes. Certain sustainability issues are inherent in cultivating and producing conventional natural fibres. In the case of cotton, the increasing cotton productivity owing to the use of vast amounts of chemicals has caused various environmental impacts. Certain problems associated with conventional cotton are water pollution, decreased levels of soil fertility, biodiversity loss, resistance to pesticides, and health issues for the workers [4].

The water usage required to produce 1 kg of cotton in India is exceptionally high, as reported by the Water Footprint Network. It takes approximately 22,500 L of water to cultivate this quantity of cotton, which is significantly more than the global average of 10,000 L. This discrepancy can primarily be attributed to the inefficient utilization of water and the high prevalence of water pollution caused by the extensive use of chemical pesticides [8, 9]. As per the statistics, traditional cotton cultivation consumes 4.7% of the total pesticides against only 2.4% of the total land. Approximately, in developing countries, almost 50% of the fertilizers are used in cotton cultivation alone [10, 11].

Wool is the only fibre that requires wet cleaning ahead of yarn manufacturing as raw wool contains many impurities containing dirt and grease. However, some chemical solvents are also used to replace the use of water. During the pretreatment stage of wool, an effluent is produced, i.e., wool grease sludge which is high in suspended solids content with a high pollution index [4].

As per the data available, polyester fibre production contributed to greenhouse gas emissions, in 2022 alone 125 MT of carbon dioxide was produced out of 47 MT of fibre production [12].

3 Introduction to Sustainable Fibres

Owing to the increased interest in sustainability, different natural fibres are preferred for developing biodegradable textile products [13]. Different natural fibres such as jute, cotton, silk, ramie, wool, flax etc. and regenerated fibres like viscose, etc., are invariably claimed to be sustainable fibres due to their biodegradability [14, 15]. Moreover, extensive studies have also been executed to explore the potential applications of easily available agricultural plant waste like leaf and bark for the extraction of textile-grade fibres for textile utility [16–19].

Considering the issues mentioned before, the following sustainable alternatives are being favoured over conventional natural fibres.

3.1 Organic Cotton, Naturally Coloured Cotton, Kala Cotton

In the case of cotton cultivation, the biggest challenge concerning sustainability lies in reducing chemical use and water consumption. The major solution to address all these issues suggests not using any of these resources and growing cotton organically while safeguarding the interest of the farmers [4].

In recent years, many brands have promoted the use of organic cotton in varying proportions. As per the Textile Exchange report, 2022 global organic cotton production has increased consecutively for 5 years. Increasing demand for organic cotton has been the prime factor for the growth of organic cotton production. It is very encouraging to the existing farmers and also allows non-organic farmers to get inclined towards using organic production methods [20].

As per the report, India contributes to the largest share of production of organic cotton, i.e., 51%, which is popularly used as a sustainable textile. The production of organic textile is expected to grow at a compound annual growth rate of 40.0% in the 2021–28 period, reaching the market size of US$ 6.73 billion [21]. Organic cotton production benefits the environment by conserving water, improving soil quality, reducing dependence on harmful chemicals, enhancing biodiversity, potentially lowering greenhouse gas emissions, and improving the health of small-scale farmers [22].

Another sustainable cotton variety is naturally coloured cotton, which is grown in different shades of beige, brown etc. It does not negatively impact the environment as available naturally coloured fibres are used for further processing without dyeing.

Naturally coloured fibres show colour variations prominently in fibres like cotton and wool. Cultivating a coloured fibre that is not processed using chemicals avoids the impacts arising out of dyeing and printing. In earlier days, 8–12 shades of reds, browns, beige and greens of native cotton were conserved by traditional people. Using naturally coloured cotton fiber as textile could reinforce sustainable practices in the textile and fashion industry as it leads to water, energy and chemical savings by reducing the need for dyeing [23–25]. Conventional breeding is also followed to produce naturally coloured cotton and one such variety is Pima sutra cotton. This variety has a range of shades of browns, greens, and reds. Different naturally coloured cotton varieties are selectively bred to achieve noticeable colours with superior fibre quality [26]. Different techniques such as agronomic practices, molecular genetics and genetic crossbreeding are used to develop naturally coloured cotton [27]. As per one study, considering the satisfactory air permeability and mechanical properties of naturally coloured organic Turkish cotton textile, fabrics were recommended for making apparel garments [28].

Kala cotton, an indigenous variety of cotton grown in Gujrat, is one of the best sustainable alternatives to conventional cotton. It is purely a rain-fed crop and can grow in arid regions. An NGO situated in Kutch, Gujrat, which works towards the preservation and proliferation of Kala cotton claims that only about 0.11% of GHG emissions are generated by Kala cotton as compared to 0.63% of GHG emissions by conventionally irrigated cotton [29].

3.2 *Organic Wool*

Organic wool is obtained from sheep that graze on pastureland and is not treated with pesticides or reared on organically grown feed and forage. Moreover, any kind of induction of synthetic hormones and genetic modification on wool is not permitted. The sheep must not be treated with parasiticides; hence, using organic wool is good for both animal health and soil [30]. Harsh chemicals are used in scouring and dyeing conventional wool, resulting in an increased environmental impact. Hence organic wool shall be pretreated and processed in an ecofriendly manner to ensure its minimal environmental impact [31].

A kind of sustainability standard, the Responsible Wool Standard, i.e., RWS, was launched by Textile Exchange in 2016. It aims to help promote traceability for responsible wool produced with improved environmental, social, and animal welfare practices [32]. As per different studies, consumers are buying organic, eco-friendly, natural, and sustainable products at higher price points as compared to conventional ones, with organic being the most preferred option [33, 34].

3.3 Organic Silk

Unlike mulberry silk, organic silk production is eco-friendly, nonviolent, and sustainable. In its production, silkworms are allowed to have natural death after living their full life. The silkworm becomes a moth after living a natural lifecycle. Once the moth emerges naturally, the silkworm chrysalis is collected; hence, peace/vegetarian silk is used to refer to this kind of silk obtained from wild silkworms as no violence is used throughout the process. These silk filaments show various textures and colours in comparison to conventional mulberry silk. The length of such fibres is comparatively lesser which renders a raw hand feel to the spun yarns.

A type of organic silk is wild peace silk, where silkworms are cultivated in an open forest without using any hazardous chemicals. In countries like India, China and Vietnam, insects belonging to the mainly *Antherea* family living in wild or semi-domesticated and thus produced silk is commonly known as Tassar. Instead of mulberry leaves, these insects survive on oak leaves, especially in areas where small oak trees grow mostly in Himalayan foothills [35].

During the collection process of cocoons and converting the fibres into fabric by manual methods, employment is generated for the local villagers and Indigenous people. In central and eastern Indian plateaus, sericulture is preferred as a livelihood option by a large number of poor families [36]. Wild silk is earth-friendly and can encourage forest preservation that is otherwise threatened by deforestation. Two other silk varieties, i.e., Asan and Arjun, are also produced in India. Muga fibre is gold in colour and often discontinuous, while eri silk is obtained from a variety of silkworms that survive on castor leaves. Hence, organic silk cultivation ensures environmental, social and economic sustainability. Various blends of raw silk in combination with cotton and any other fibres are made to add sheen and softness [37].

3.4 Organic Linen

Various weed control methods are used in the cultivation of linen, especially the use of fertilizers and pesticides; however, if plants are sufficiently irrigated, there may not be any requirement for fertilizers. The moist climate is suitable for achieving superior-quality fibres. Bast fibres like linen can grow well on land rendered unsuitable for food crop cultivation, and it is also reported to reduce heavy metal contamination [38].

Fibre extraction methods such as dew retting and enzyme retting are considered sustainable extraction techniques that reduce environmental pollution levels as compared to the chemical retting method [39].

3.5 Hemp

Hemp fibres are extracted from stalks of hemp plants. These can effectively grow without pesticides and herbicides, giving high yield with following a low labour-intensive production process. Hemp fibre is a sustainable, environmentally friendly, and versatile textile fibre. It is being proposed as eco-friendly alternative to conventional cotton and synthetic fibres used in the textile industry [40, 41].

Various products like textiles, ropes, mulching materials, animal bedding, etc., are made of hemp fibres. Hemp is also used to produce a range of apparel like casuals, workwear, bed linen, accessories, carpets and drapes etc. For apparel, the blends of hemp with other fibres give a better hand feel, drapability like silk, cotton, Tencel and lyocell, etc. These fibres possess various favourable attributes like breathability, durability, UV resistance, antimicrobial, and resistant to mold and mildew. It readily dyes and softens with each consecutive wash cycle. It is similar to linen in texture [42]. Recently hemp is also being increasingly used in wipes and hygiene products.

3.6 Lotus Fibres

These are obtained from the stem/stalk of the lotus plant. Lotus fibre has a low density and good absorbency, similar to silk and acrylic, but with a higher moisture regain than cotton and silk [43]. Lotus fibres are spun and woven by hand, employing traditional techniques.

Textiles made of lotus fibres are naturally extremely soft, light and anti-bacterial [44]. Fabric made out of lotus fibres is environment-friendly organic fabric, and it has the potential to revolutionize the luxury sector due to its wrinkle-resistant, breathable properties and purity [45].

3.7 Pineapple Fibre

After harvesting the pineapple fruit, the leaves of the pineapple plant remain unutilized. The waste pineapple leaves are used to obtain the Piña fibres. The strength, lightness, and shiny appearance of pineapple leaf fibre make it a valuable fibre. Pineapple fibres are blended with silk and cotton fibres, such as silk or cotton, to enhance their qualities and create unique textile fabrics [46].

Indian pineapple leaf fibre (PALF) can be efficiently extracted, processed, and converted into sustainable novelty fabrics for apparel and novelty textiles [47]. Moreover, it offers valuable properties for making dress material, accessories and home furnishing products [48]. Pineapple leaf fibre can be dyed with natural dyes [49]. Pineapple leaf fibre (PALF) shows potential as an eco-friendly, biodegradable, and renewable raw material for various applications; however, a detailed study of its properties is needed for logical and reasonable utilization [50].

3.8 *Bamboo*

Bamboo is organically grown without needing any pesticides and fertilizers. It is made up of cellulose that is obtained from fast-growing and woody bamboo grass. Two types of bamboo fibres are used in the industry; one is natural bamboo, also referred to as bamboo linen due to its similar hand feel to linen. Natural bamboo is extracted from bamboo culms. The other type of bamboo is bamboo viscose, which is more commonly known in the industry. In this case, bamboo is used as raw cellulose and it goes through a similar process as viscose production. There are two types of chemical processes used; in the first one, the viscose spinning method is used, and in the second one, a closed-loop solvent system is followed, as in the production of Lyocell fibres [51]. Such fibres are termed as Lyocell itself [52].

Natural bamboo fibres are usually less used in the manufacturing of apparel. Generally, 3- to 4-year-old bamboo plant is used for fibre extraction. Several steps are used like alkaline hydrolysis, multiphase bleaching, and chemical treatment of starchy pulp. Owing to the micro gaps in bamboo fibres, these are softer than cotton, with better moisture absorption. These fibres are eco-friendly and biodegradable. The various positive attributes of bamboo fibres, like antifungal, durability, bacteriostatic, UV resistance, etc., make it suitable for various end uses like apparel, bath linen etc. It is also used for hygiene products, medical PPE kit, etc. [53].

Bamboo fibres can be used to create green, environmentally responsible eco-products, with processing parameters affecting their performance and functionality in various industrial applications. Regenerated bamboo fibre shows potential in the textile industry, with its strengths and weaknesses compared to viscose rayon and carded and combed ring-spun yarns [54].

3.9 *Banana Fiber*

It is derived from the pseudo-stem part of the plant. It offers a sustainable, biodegradable and lightweight alternative to synthetic fibres for various industrial applications including sanitary pads, textiles, and bio-based products [55, 56]. Enzyme and chemical combined treatment significantly improve the fineness of banana fibres, enabling their use as textile and packaging materials [57, 58].

4 Case Studies of Sustainable Textile Brands and Designers

Today, textile companies are aware of the potentially harmful effects of their actions on the environment and work within strict guidelines concerning ecology. Now, they are more open to adopting effective measures to reduce their detrimental impact on the environment. However, many brands and companies are misleading consumers by greenwashing. Hence a lot of ambiguity prevails in the industry concerning using terms like ethical, sustainable, ecofriendly, green, and climate positive etc. Nonetheless, some companies and brands have taken the onus on themselves for a climate-positive future and are trying hard to approach sustainability from various perspectives. Various textile manufacturing companies are making efforts to adopt sustainable practices; however, there is a long way to go.

Some case studies of such textile stakeholders are presented below:

4.1 Pratibha Syntex

Pratibha is India's largest sustainable apparel manufacturer and exporter, offering a comprehensive fibre-to-fashion solution. The company was founded in 1997 by Mr. Shiv Kumar Chaudhary. It is a globally sourced, responsible garment manufacturer because of its commitment to ethical practices and environmental sustainability. Different types of product ranges are crafted in Pratibha by the name of Vasudha Primo, Allure, Loop and Xterra (Fig. 1).

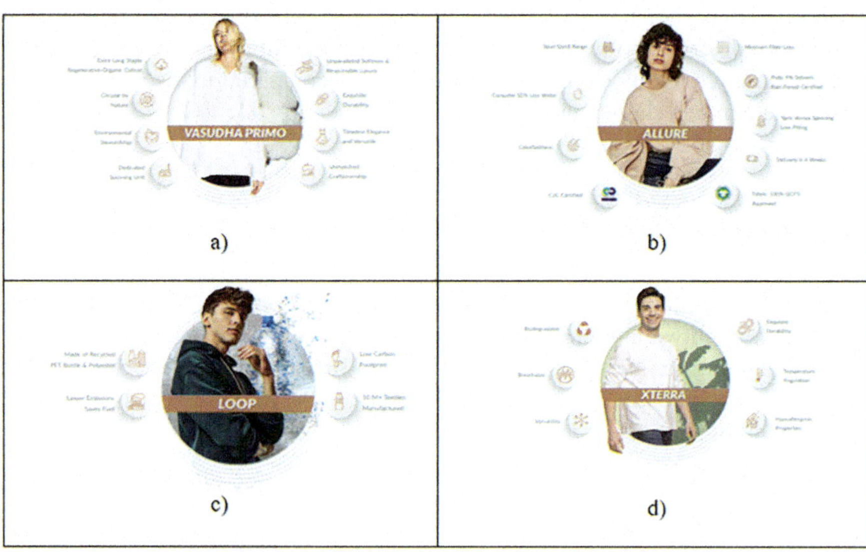

Fig. 1 Attributes of different ranges of Pratibha Syntex (**a**)Vasudha Primo (**b**) Allure (c) Loop (d) Xterra [59]

Vasudha Primo range as shown in Fig. 1 (a) combines regenerative and carbon-neutral materials with water-efficient processes by using extra-long-staple organic & regenerative organic cotton. The resulting range is soft, durable and versatile, aligning fashion with nature and social responsibility.

Its Allure range (b) of products is a breakthrough in sustainability that utilizes waterless dyeing techniques. It showcases vibrant, eco-friendly colours while conserving precious water resources and makes a fashionable statement with a responsible impact. The company's Loop range is created entirely from recycled materials which transform plastic waste into eco-conscious attire. It shows their commitment to sustainability, innovation, and a greener fashion. The Xterra range (c) constitutes finely crafted, eco-conscious knitted garments, which are made from sustainable hemp, luxurious silk and breathable linen [59].

4.2 *Stella McCartney*

The eponymous luxury fashion brand was launched in 2001. The brand has never used any animal byproducts, like skin, fur, leather, etc., in any of its collections. The designer also collaborates with many innovative companies. The brand is known for developing sustainable innovative materials by continuously improving the sustainability parameters of products. The brand uses only regenerative cashmere to reduce the environmental impact. The brand also exhibited Future of Fashion material innovation at UNCCC (COP26). Wood is obtained from sustainably managed forests for viscose production. It has collaborated with the innovative textile company Bolt Threads to promote the development of sustainable materials. The brand also launched ECONYL® based close-the-loop garment, ensuring a circular design process. Reinforcing the development of innovative materials, Mycelium was used by the brand to create the Mylo™ luxury handbag. The garments made out of BioSequin® and regenerative cotton were firstly produced by the brand itself. Stella McCartney made a sustainable eveningwear collection named Green Carpet Collection. The brand is known to use regenerated cashmere to lessen the negative impact. The brand continues to make progress in the direction of sustainable textile and fashion by collaborating with sustainability-conscious companies and brands [60].

4.3 *No Nasties*

No Nasties was founded by Apurva Kothari in 2011 in Goa, India. It makes 100% organic, fair trade, vegan clothing using a local supply chain. The brand claims to offset the carbon footprint by using solar power, wind energy, and reforestation. For every product bought, three trees are planted. They have certain other features to ensure the sustainable use of their clothes. The almost new clothes could be sold again on its website directly to customers via its Relove program. Garments with minor tears or stains could be sent back to the company to get them repaired or can be donated to a local thrift store. The very old clothes are converted back to yarns and new products are developed. They work with factories that do not support child labour and worker exploitation. They have also created a 3-year roadmap in which they propose to use more natural dyes, making clothes 100% biodegradable and generating zero waste in the production process of garments. They also propose to use recycled & even regenerative cotton, using compostable packaging and renewable energy. Their focus is also third-party verification of climate positivity for better visibility of their climate actions [61].

4.4 Ananas Anam

It is a for-profit textile company that claims to have less environmental impact and a positive social impact. It considers developing innovative products while promoting environmental, social and cultural welfare. The Founder and Chief Creative & Innovation Officer, Dr. Carmen Hijosa, for Ananas Anam Ltd., envisions a future that sustainably connects people, the planet and profit. The company states that every year, a huge quantity of waste is generated from pineapple agriculture globally. The leaves of the pineapple plant are traditionally discarded or burned. Piñatex®, an innovative material, was produced to be used as an alternative to bulk-produced leather and synthetic materials. The production of Piñatex® assures no use of extra natural resources and no banned chemicals on the list of Cradle2Cradle. Different collections are made using Piñafelt, which is dyed using GOTS-certified pigments. Additional durability, strength and water resistance are achieved by applying a top coating of resin. The metallic collection has been created by heat pressing a foil, and another collection of Piñatex Performance Collection is developed using a high solid PU transfer coating.

Piñatex material is suitable for use in the manufacturing of fashion, accessories and upholstery. It has been used by more than 1000 brands globally, like H&M, Hugo Boss, and Hilton Bankside. As per the claims made by the company, every meter of it reduces the emission of carbon dioxide, and per kg of Piñayarn reduces the emission of 6 kg of CO_2. Figure 2 shows various applications of Pineapple fibre and lifecycle stages of the PAF [62].

4.5 Bolt Threads

The company was founded in 2009 in the United States by a bioengineer, a biophysicist, and a biochemist. A revolutionary material, Mylo, was engineered with hand feel similar to that of leather. It is soft, supple and not harmful to the environment. It is a sustainable leather alternative made from mycelium, the root-like

Fig. 2 The manufacturing and lifecycle of PIÑATEX [62]

system of mushrooms grown in the lab setting. Mylo™ successfully passed rigorous testing for the presence of over 350 harmful and toxic chemicals, earning OekoTex Standard 100. The materials having robust controls on quality and safety also meet Bluesign® certification. Mylo is also vegan-verified by Eurofins I Chem-MAP. Another innovative sustainable material developed by Bolt Threads is Microsilk, which is produced with less environmental impact than traditional textile manufacturing. The company develops proteins inspired by these natural silks by using bioengineering to put genes into yeast; then, they produce the protein in large quantities through fermentation, using yeast, sugar, and water before finally knitting fibres into fabrics and garments. A shift dress made with Microsilk fabric was made in collaboration with designer Stella McCartney at the New York MoMA in October 2017. They also developed a limited edition iconic knit hat made from a blend of rambouillet wool and Microsilk fibre in collaboration with Best Made Co. The Biofabric Tennis Dress, created with a Microsilk and cellulose blend fibre, was manufactured by Stella McCartney and Adidas in 2019 [63].

4.6 Patagonia

Patagonia is an outdoor and casualwear clothing brand, and it is known for its sustainability and environmental activism. The marketing strategies and business practices of the company are designed to minimize environmental damage and promote sustainability. Patagonia's approach to eco-design, which includes the use of organic, recycled, and traceable materials, is a significant aspect of its product line development, reflecting a focus on functionality over aesthetics. Interestingly, despite Patagonia's efforts to distance itself from the fashion industry, its practices contribute to increased consumption, which can be seen as a paradox in its green marketing approach. The company's success and influence are evident in its high ranking in the Fashion Transparency Index and its international presence, which underscores its role in shaping the fashion industry's approach to sustainability. While the brand has made significant strides in promoting eco-friendly practices and materials, it also faces the challenge of reconciling its environmental ethos with the realities of consumer behaviour and market demands. Patagonia's story serves as a valuable example for other companies seeking to navigate the intricate balance between sustainability, consumerism, and business growth [64–66].

Some other well-known sustainable brands are Eileen Fisher, PeopleTree, The Hemp clothing trading company, etc., are striving to follow sustainability practices in their production system.

5 Impact of Sustainable Practices on Design

The design represents the crucial step in ensuring sustainable practices in the whole supply chain. Various methods to slow consumerism include the use of organic fabrics, high-quality fabrics, durable products etc.

The rising level of textile waste and overflowing landfill sites lead the designers to design and use fibres and materials which are produced with reduced waste and biodegradable and recyclable fibres and which are certified Fairtrade fibres [4].

Keeping sustainability in mind, designers find ways to produce textiles with increased lifespan. Sustainability advocates a holistic Cradle-to-grave approach which considers the life cycle assessment (LCA) of textiles. As a result, new production methods are devised to reduce the impact of the industry's carbon footprint and decrease textile waste. Today textile designers are understanding sustainability in different ways. There is an increasing trend toward using recycled materials, organic fabrics, and biodegradable renewable materials. Some designers are also emphasizing socially responsible methods that create more job opportunities for marginalized workers to boost local economies [67]. Sustainability is compelling designers to use renewable fibres and use materials that require less energy, water and chemicals.

Table 1 illustrates the various strategies for sustainable design. Textile designer Luisa Cevese uses industrial waste and transforms it into materials that are used to create very desirable fashion accessories and textiles for exhibitions [69].

Discarded clothing can be upcycled by adding value to the products and recycling of fabrics can be done to prevent textile waste. It is very crucial for designers to be aware of the environmental implications associated with making textile products out of fibres. Since various stages of manufacturing demand the consumption of resources and energy, which imposes an impact on the environment. Hence, a holistic integrated approach to design and manufacture of products is necessary.

Despite the efforts made by various industry stakeholders, the challenge remains the same, i.e., to provide diverse, accessible products to clients that integrate sustainability while considering the aesthetic aspect of design. This further paves the way for new avenues and new dimensions of textile design, with sustainability being the prime one [70].

Table 1 Ten sustainable design strategies [68]

S. no.	Sustainable design strategy	Example
1	Design for minimum waste generation	Recycling, zero-waste manufacturing
2	Design for recycling/upcycling	Circular design systems
3	Design for minimum chemical impacts	Using the materials and processes
4	Design for minimum water and energy usage	Zero-waste printing and dyeing, air dyeing, design for no need for laundering
5	Design that explores clean and better technology	Laser/water jet/sonic cutting, digital printing
6	Design that gets inspired by nature	Exploring biomimicry
7	Design for ethical production system	Traditional craft skills, procuring fair trade materials, value worker's rights
8	Design for reducing the need for consumption	Classic designs, slow designs, culture of DIY, customized textiles
9	Design for developing system and services to dematerialize	Lease, repair, share
10	Design for increasing awareness	Increasing awareness of designers regarding the environmental and social impacts of textile products

6 Conclusion

Conventionally textile design had been only about adding value to products by various means of weaving, printing, embroidery, embellishments, finishes, etc. However, with enhancing consumer awareness of the detrimental impact of the textile industry on the environment, there has been a great realization that for the long-term sustenance of the textile industry, the designers need to be more holistic with their approach at every stage of manufacturing and design. The primary obstacles to achieving sustainability include consumers, costs, and traditional company structures. To incorporate sustainable design principles and practices within the companies, there has to be a reinvented system for inter-departmental collaborations. Moreover, cross-departmental initiatives and actions are essential for large textile businesses, as innovation cannot be limited to design departments. Hence a comprehensive approach is necessary for embedding sustainability in textile design, requiring the coordination of multiple stakeholders, including fibre cultivators, textile manufacturers, distributors, and consumers. By taking deliberate actions, these parties can ensure the sustainability of the final product and its long-term advantages. The widespread adoption of sustainable textiles can significantly contribute to the future of the textile industry and the environment.

References

1. Hethorn, J., & Ulasewicz, C. (2008). *Sustainable fashion: Why now? A conversation exploring issues, practices, and possibilities*. Bloomsbury Academic.
2. Fletcher, K., & Grose, L. (2012). *Fashion and sustainability*. Laurence King Publishing.
3. Textile Exchange. (2023). *Report*. https://textileexchange.org/knowledge-center/reports/materials-market-report-2023/
4. Fletcher, K. (2008). *Sustainable fashion and textiles*. Earthscan.
5. Chattopadhyay, S. N., Pan, N. C., Roy, A. N., Samanta, K. K., & Khan, A. (2021). Hybrid bleaching of jute yarn using hydrogen peroxide and peracetic acid. *Indian Journal of Fibre and Textile Research, 46*, 78–82.
6. Debnath, S. (2016). Natural fibres for sustainable development in fashion industry. In S. Muthu & M. Gardetti (Eds.), *Sustainable fibres for fashion industry. Environmental footprints and eco-design of products and processes* (pp. 89–108). Springer. https://doi.org/10.1007/978-981-10-0522-04
7. Jagadeesh, P., Puttegowda, M., Girijappa, Y. G. T., Rangappa, S. M., & Siengchin, S. (2022). Effect of natural filler materials on fiber reinforced hybrid polymer composites: An overview. *Journal of Natural Fibers, 19*(11), 4132–4147. https://doi.org/10.1080/15440478.2020.1854145
8. Leahy, S. (2015). World water day: The cost of cotton in water-challenged India. Guardian Sustainable Business. *The Guardian*. https://www.theguardian.com/sustainable-business/2015/mar/20/cost-cotton-water-challenged-india-world-water-day
9. Water Footprint Network. (n.d.). World water day: The cost of cotton in water-challenged India. *The Guardian*. https://www.theguardian.com/sustainable-business/2015/mar/20/cost-cotton-water-challenged-india-world-water-day#:~:text=Producing%201kg%20of%20cotton%20in,is%20too%20contaminated%20for%20reuse
10. Better Cotton. (2023). *Pesticides and crop protection*. https://bettercotton.org/field-level-results-impact/key-sustainability-issues/pesticides-and-crop-protection-in-cotton-farming/
11. Scheer, M. R. (2001). A sustainable vision on the cotton industry after 2005. In *Proceedings of the Speech Held at the General Assembly of EUROCOTON, Brussels, Belgium, 30 May 2001*
12. Textile Exchange. (2024). *Report on the future of synthetics*. https://textileexchange.org/knowledge-center/reports/the-future-of-synthetics/
13. Duarte, L. O., Kohan, L., Pinheiro, L., Fonseca Filho, H., & Baruque-Ramos, J. (2019). Textile natural fibers production regarding the agroforestry approach. *SN Applied Sciences, 1*, 914.
14. Nayak, R. (2019). *Sustainable technologies for fashion and textiles*. Woodhead Publishing.
15. Thomas, S., Paul, S. A., Pothan, L. A., & Deepa, B. (2011). Natural fibres: Structure, properties and applications. In *Cellulose fibers: Bio-and nano-polymer composites*. Springer.
16. Chattopadhyay, S. N., Pan, N. C., Roy, A. N., & Samanta, K. K. (2020). Pretreatment of jute and banana fibre: Its effect on blended yarn and fabric. *Journal of Natural Fibres, 17*(1), 75–83. https://doi.org/10.1080/15440478.2018.1469450
17. Gupta, P. K., Patra, S., & Samanta, K. K. (2021). Potential of okra for application in textiles: A review. *Journal of Natural Fibres, 18*(11), 1788–1800. https://doi.org/10.1080/15440478.2019.1697997
18. Kambli, N. D., Samanta, K. K., Basak, S., Chattopadhyay, S. K., Patil, P. G., & Deshmukh, R. R. (2018). Characterization of the corn husk fibre and improvement in its thermal stability by banana pseudostem sap. *Cellulose, 25*(9), 5241–5257. https://doi.org/10.1007/s10570-018-1931-z
19. Sunny, G., & Rajan, T. P. (2021). Review on areca nut fiber and its implementation in sustainable products development. *Journal of Natural Fibers, 19*(12), 1–15. https://doi.org/10.1080/15440478.2020.1870623
20. Textile Exchange. (2022). *Organic cotton market report*. https://textileexchange.org/knowledge-center/reports/organic-cotton-market-report/

21. IBEF. (2024). *A green sustainable cotton: Kala cotton*. https://www.ibef.org/research/case-study/a-green-sustainable-cotton-kala-cotton
22. Bachmann, F. (2011). Potential and limitations of organic and fair-trade cotton for improving livelihoods of smallholders: Evidence from Central Asia. *Renewable Agriculture and Food Systems, 27*, 138–147. https://doi.org/10.1017/S1742170511000202
23. Chaudhry, M. R. (1992). Natural colours of cotton. *ICAC Recorder, Washington, 10*(4), 3–5.
24. Garciai, S., & de Alencar, I. (2014). Textile industry can be less pollutant: Introducing naturally coloured cotton. *International Journal of Product Management Engineering, 2*(2), 85–91.
25. Günaydin, G., Avinc, O., Palamutçu, S., Yavaş, A., & Soydan, A. (2018). Naturally colored organic cotton and naturally colored cotton fiber production. In *Organic cotton*. Springer. https://doi.org/10.1007/978-981-10-8782-0_4
26. Sun, J., Sun, Y., & Zhu, Q. H. (2021). Breeding next-generation naturally coloured cotton. *Trends in Plant Science, 26*(6), 539–542.
27. Madhu, A. (2024). *Naturally Colored cotton: A sustainable innovation*. IntechOpen. https://doi.org/10.5772/intechopen.113290
28. Günaydina, G., Palamutçu, S., Soydan, A., Yavaş, A., Avinc, O., & Demirtaş, M. (2020). Evaluation of fiber, yarn, and woven fabric properties of naturally colored and white Turkish organic cotton. *The Journal of the Textile Institute, India, 111*, 1436–1453. https://doi.org/10.1080/00405000.2019.1702611
29. Jha, B. (2018). Kala cotton: A sustainable alternative. In *The Asian Conference on Sustainability, Energy & the Environment 2018 Official Conference Proceedings*. The International Academic Forum
30. Anonymous. (2009). *What does organic wool mean?* https://oecotextiles.blog/2009/08/11/what-does-organic-wool-mean/
31. Miller, J. (2023). *What is organic wool-is it really ethical and sustainable*. https://theroundup.org/what-is-organic-wool-ethical-sustainable/
32. Textile Exchange. (2024). *Report*. https://textileexchange.org/app/uploads/2024/03/Textile-Exchange-Annual-Report-2023.pdf
33. Hustvedt, G., & Dickson, M. (2009). Consumer likelihood of purchasing organic cotton apparel: Influence of attitudes and self-identity. *Journal of Fashion Marketing and Management, 13*, 49–65. https://doi.org/10.1108/13612020910939879
34. Bernard, J., Hustvedt, G., & Carroll, K. (2013). What is a label worth? Defining the alternatives to organic for US wool producers. *Journal of Fashion Marketing and Management, 17*, 266–279. https://doi.org/10.1108/JFMM-01-2013-0009
35. Franck, R. R. (2001). *Silk, mohair, cashmere and other luxury fibres*. Woodhead Publishing.
36. Professional Assistance for Development Action (PRADAN). (n.d.). *Report*. https://www.pradan.net//images/documentation/SilkenSpread_Pradan.pdf
37. Karthik, T., Rathinamoorthy, R., & Ganesan, P. (2015). Sustainable luxury natural Fibers: Production, properties, and prospects. In M. Gardetti & S. Muthu (Eds.), *Handbook of sustainable luxury textiles and fashion. Environmental footprints and eco-design of products and processes*. Springer. https://doi.org/10.1007/978-981-287-633-1_4
38. Eyupoglu, S. (2020). Sustainable plant-based natural fibers. In S. Muthu & M. Gardetti (Eds.), *Sustainability in the textile and apparel industries. Sustainable textiles: Production, processing, manufacturing & chemistry*. Springer. https://doi.org/10.1007/978-3-030-38541-5_2
39. Tahir, M. P., Ahmed, A., SaifulAzry, S., & Ahmed, Z. (2011). Retting process of some bast plant fibres and its effect on fibre quality: A review. *BioResources, 6*, 5260–5281. https://doi.org/10.15376/biores.6.4.5260-5281
40. Malabadi, R., Kolkar, K., & Chalannavar, R. (2023). Industrial Cannabis sativa: Role of hemp (fiber type) in textile industries. *World Journal of Biology Pharmacy and Health Sciences*. https://doi.org/10.30574/wjbphs.2023.16.2.0450
41. Gedik, G., & Avinc, O. (2020). Hemp fiber as a sustainable raw material source for textile industry: Can we use its potential for more eco-friendly production? In *Sustainability in the textile and apparel industries* (pp. 87–109). Springer. https://doi.org/10.1007/978-3-030-38541-5_4

42. Kaur, G., & Kander, R. (2023). The sustainability of industrial hemp: A literature review of its economic, environmental, and social sustainability. *Sustainability, 15*(8), 6457. https://doi.org/10.3390/su15086457
43. Mahapatra, C. (2012). *Processing of lotus fiber in textile industries* (Vol 2012, pp. 57–59). Colorage.
44. Samatoa Lotus Textiles. (n.d.). *Sustainable fabric, samatoa lotus textiles, sustainable production*. https://samatoa.lotus-flower-fabric.com/
45. Gardetti, M., & Muthu, S. (2015). The lotus flower fiber and sustainable luxury. In *Handbook of sustainable luxury textiles and fashion* (pp. 3–18). Springer. https://doi.org/10.1007/978-981-287-633-1_1
46. Jose, S., Das, R., Mustafa, I., Karmakar, S., & Basu, G. (2019). Potentiality of Indian pineapple leaf fiber for apparels. *Journal of Natural Fibers, 16*, 536–544. https://doi.org/10.1080/15440478.2018.1428844
47. Hazarika, D., Gogoi, N., Jose, S., Das, R., & Basu, G. (2017). Exploration of future prospects of Indian pineapple leaf, an agro waste for textile application. *Journal of Cleaner Production, 141*, 580–586. https://doi.org/10.1016/J.JCLEPRO.2016.09.092
48. Rao, A. (2017). A different approach towards textile by pineapple fiber. *Asian Journal of Multidisciplinary Studies, 5*.
49. Shukla, P., Ghosh, H., & Joshi, K. (2017). Dyeing pineapple fiber with Bixa orellana (Achiote) natural dye. *International Journal of Applied Research, 3*, 1126–1128.
50. Asim, M., Abdan, K., Jawaid, M., Nasir, M., Dashtizadeh, Z., Ishak, M., & Hoque, M. (2015). A review on pineapple leaves fibre and its composites. *International Journal of Polymer Science, 2015*, 950567. https://doi.org/10.1155/2015/950567
51. Nayak, L., & Mishra, S. (2016). Prospect of bamboo as a renewable textile fiber, historical overview, labeling, controversies and regulation. *Fashion and Textiles, 3*, 1–23. https://doi.org/10.1186/s40691-015-0054-5
52. Imadi, S., Mahmood, I., & Kazi, A. (2014). Bamboo fiber processing, properties, and applications. In K. Hakeem, M. Jawaid, & U. Rashid (Eds.), *Biomass and bioenergy* (pp. 27–46). Springer. https://doi.org/10.1007/978-3-319-07641-6_2
53. Liu, D., Song, J., Anderson, D., Chang, P., & Hua, Y. (2012). *Bamboo fiber and its reinforced composites: Structure and properties* (Vol. 19, pp. 1449–1480). Cellulose. https://doi.org/10.1007/s10570-012-9741-1
54. Erdumlu, N., & Ozipek, B. (2008). *Investigation of regenerated bamboo fibre and yarn characteristics*. Fibres & Textiles in Eastern Europe.
55. Balda, S., Sharma, A., Capalash, N., & Sharma, P. (2021). Banana fibre: A natural and sustainable bioresource for eco-friendly applications. *Clean Technologies and Environmental Policy, 23*, 1389–1401. https://doi.org/10.1007/s10098-021-02041-y
56. Iqbal, S., Kabir, M., Muhib, M., & Rupok, M. (2023). Prospect of waste banana fiber use in industry: A narrative review. *International Journal of Science and Research Archive, 10*(2), 828–839. https://doi.org/10.30574/ijsra.2023.10.2.1045
57. Balakrishnan, S., Wickramasinghe, G., & Wijayapala, U. (2019). Investigation on improving banana fiber fineness for textile application. *Textile Research Journal, 89*, 4398–4409. https://doi.org/10.1177/0040517519835758
58. Vigneswaran, C., Pavithra, V., Gayathri, V., & Mythili, K. (2015). Banana fiber: Scope and value added product development. *Journal of Textile and Apparel Technology and Management, 9*, 1–7.
59. Pratibha Syntex. (n.d.). *More than responsible textiles*. https://pratibhasyntex.com/
60. Stella McCartney US. (n.d.). *Designer RTW, Bags & accessories, Lingerie, Adidas by Stella McCartney, Fragrances, Kids*. www.stellamccartney.com
61. No Nasties. (n.d.). *100% organic fair trade vegan cotton clothing*. https://www.nonasties.in/
62. Ananas Anam. (2024, January 8). *Home*. https://www.ananas-anam.com
63. Bolt Threads. (n.d.). https://boltthreads.com/

64. Hepburn, S. J. (2013). In patagonia (clothing): A complicated greenness. *Fashion Theory, 17*(5), 623–645. https://doi.org/10.2752/175174113X13718320331035
65. Patagonia UK. (n.d.). *Outdoor clothing & gear*. https://www.patagonia.com/
66. Wang, L., & Shen, B. (2017). A product line analysis for eco-designed fashion products: Evidence from an outdoor sportswear brand. *Sustainability, 9*(7), 1136. https://doi.org/10.3390/su9071136
67. Quinn, B. (2013). *Textile visionaries: Innovation and sustainability in textile design*. Laurence King Publishing.
68. Black, S. (2012). *The sustainable fashion handbook*. Thames and Hudson.
69. Clarke, S. E. B., & O'Mahony, M. (2006). *Techno textiles 2: Revolutionary fabrics for fashion and design*. Thames and Hudson.
70. Gordon, J. F., & Hill, C. (2015). *Sustainable fashion past, present and future*. Bloomsbury Publication.

Technological Advancements in Textile Design

M. Pavan, Kshetrimayum Dolly Devi, N. Sowmya, Saikhom Debina Chanu, and Surabhi Mahajan

1 Introduction

Rapid technological advances in recent years have brought about substantial changes in the textile sector. The evolution from manual textile production to the current age of digitalization and automation has significantly transformed design, manufacturing, and product development processes [1, 2].

The textile industry has witnessed a notable technological breakthrough with the implementation of computer-aided design (CAD) software. The way that designers construct clothing and textiles has changed as a result of tools like Optitex, Browzwear, and Adobe Illustrator [3, 4]. By simulating fabric drapes, creating complex patterns, and visualizing the finished product virtually, these software platforms speed up design development and lessen the need for physical prototypes [5].

Additionally, the emergence of 3D textile printing (3DP) technology has reformed textile design by facilitating the designing of intricate and personalized textile assemblies and garments that are difficult or challenging to accomplish using conventional production techniques [6, 7]. This technology has proven especially

M. Pavan (✉)
Department of Fashion and Apparel Design, NITTE School of Fashion Technology and Interior Design, Bengaluru, Karnataka, India
e-mail: pavan.m@nitte.edu.in; pavanmanu1212@gmail.com

K. D. Devi · S. D. Chanu · S. Mahajan
Department of Apparel and Textile Science, Punjab Agricultural University, Ludhiana, Punjab, India
e-mail: mathoithoi1424@gmail.com; debinasaikhom22@gmail.com; surabhimahajanct@pau.edu

N. Sowmya
Fashion Institute of Technology, Vellore Institute of Technology, Chennai, India
e-mail: sowmyabijju@gmail.com

Sadhna et al. (eds.), *The Art and Craft of Modern Textile Design*, Sustainable Textiles: Production, Processing, Manufacturing & Chemistry,
https://doi.org/10.1007/978-3-031-86797-2_4

valuable in developing innovative fabric patterns, textured surfaces, and even entire garments.

The incorporation of intelligent and functional components constitutes a significant advancement in textile design. The emergence of "smart textiles" has made it easier to incorporate sensors, electronics, and interactive elements into textiles, producing clothing with sophisticated functions [8, 9]. These cutting-edge fabrics can follow the wearer's vital signs, respond to changes in the environment, and even change colour and pattern in response to user input [10, 11].

The technique of designing textiles has changed dramatically with the introduction of digital textile printing. With the use of this technology, designers can now directly print digital designs onto textiles, relieving them from the limitations of conventional techniques to produce one-of-a-kind patterns. Additionally, small-batch production and on-demand manufacturing are supported by digital textile printing, meeting the rising consumer demand for personalization and customization in the fashion segment [12, 13].

Furthermore, artificial intelligence (AI) and data analytics have been included in the design process by the textile sector. Now that designers have access to big data and machine learning algorithms, they may examine customer preferences, forecast trends, and optimize their designs to better align with market demands [14, 15].

All things considered, the textile design business has undergone a revolution owing to these technological developments, which have enabled designers to create more unique, customized, and sustainable textile products. As the field of textile design continues to evolve, it is expected that these technological advancements will have a substantial impact on how the field develops in the future [16, 17].

2 Digital Printing Techniques

Digital printing technology (DPT) on textiles has transformed the industry by offering numerous advantages, like reduced environmental pollution and extended design possibilities. For instance, researchers have examined the development of novel reactive disperse dyes intended for digital printing on cotton fabrics, exhibiting improved colour saturation and stability [18]. Additionally, studies have looked at the development of double-sided DPT, focusing on models for forecasting colours in heterochromatic printing on fragile textiles like chiffon and accuracy in colour management [19].

In addition, when comparing digital and manual screen printing, the limitations of screen printing, such as poor registration, colour inconsistencies, and production constraints, are underscored. This positions digital textile printing as a superior choice for short-run textile projects in countries such as Ghana, offering greater efficiency and cost-effectiveness [20].

2.1 Digital Printing Techniques

Digital printing techniques have revolutionized the textile and fashion industry. Some of the popular techniques include [21, 22]:

(a) **Direct-to-garment (DTG) method**: One cutting-edge technique for printing straight onto cotton textiles is DTG. It is one of the ideal techniques for printing complex multi-coloured designs that demand a great degree of detail and colour integrity, it's perfect.
(b) **Sublimation printing**: It encompasses applying pigment over the fabric using a heat press machine, producing premium, fade-resistant prints without compromising the breathability of the fabric.
(c) **Reactive dye printing**: This creates long-lasting, soft-to-the-touch designs with good colour fastness using water-based dyes that chemically bond with fibres.
(d) **Discharge printing**: This is application of a fabric with a colour-removing stuff in certain areas, then reprinting it with new colours to create delicate, vintage-looking patterns that are delicate and intricate.
(e) **Digital embroidery**: By combining conventional embroidery techniques with digital technology, elaborate motifs may be precisely embroidered into fabrics, adding texture and dimension to clothes.
(f) **Pigment printing**: It is growing in popularity due to its user and environmental friendliness. Commonly used for creating custom rugs, provides an eco-friendly and versatile option. The method uses less water and water-based inks, which is in line with environmentally friendly practices.

Digital printing is transforming the textile sector with several significant benefits, improving speed and efficiency and cutting down time consumption during production. Printing with digital technology is cost-effective for both small and big batches, although traditional printing often requires huge batch production. Small companies and independent designers particularly benefit from this flexibility since it allows them to create a wide range of designs without having to incur large upfront costs. Further, digital printing can produce superior quality, more vibrant colours and intricate designs. In this circular economy, where everyone is conscious about sustainability, digital printing technologies provide better solutions for reducing waste and lowering water consumption greatly as compared to traditional methods [22].

Digital textile printing encompasses various methods such as inkjet-based technology, which allows for versatile designs on different fabrics like cotton, silk, or polyester [23]. Compared to traditional screen-printing, digital printing technology (DPT) offers higher design accuracy, colour consistency, and faster changeover times between inks, with the ability to print at a resolution of 1200 dots per inch (dpi) using dispersed or reactive inks based on the fabric type [24].

Innovations such as scanning-type textile digital printing machines and equipment offer features like automatic deviation correction, high printing quality, integrated glue spreading and removal functions, efficient belt cleaning, and increased production efficiency. These advancements cater to both small and large cloth rolls, resulting in cost savings and promoting environmental sustainability. These advancements in digital textile printing technology bid a wide array of opportunities for manufacturers to produce high-quality, customizable designs efficiently [21, 24].

3 The Textile Industry and Technology

In the domain of textile and fashion, technology is essential for fostering creativity, sustainability, and productivity. The evolution of fabric manufacturing techniques has led to the aim of reducing energy utilization, environmentally friendly growth, and environmental pollution [25]. Technological innovation influences the competitive equilibrium, economic growth, and technological diffusion within the industry, creating fluctuations [26]. The amalgamation of technologies in Industry 4.0 like Big Data, IoT, blockchain and so on, in the textile supply chain, enables data-driven decision-making, traceability, and secure information recording, ensuring sustainability and facilitating a closed-loop supply chain process [27]. Furthermore, nanotechnology has revolutionized textiles, introducing smart textiles with features like antibacterial properties, pressure pads, fabric touchpads, and fabrics with phase change materials for temperature regulation [28].

3.1 Impact of Technology on Textile Sector

The notion of fashion has undergone significant evolution, heavily influenced by technological advancements that shape clothing style, design, and production. Clothing serves both functional and expressive purposes, showcasing individuality. Modern technological breakthroughs have facilitated the manufacture of synthetic fabrics tailored for diverse applications [29].

3.1.1 Mechanization in Production

Mechanization is pivotal in enhancing efficiency and quality within textile production. Numerous inventions and advancements have emerged to refine various aspects of manufacturing processes. For instance, innovations in cutting equipment focus on automating the cutting process to improve efficiency and enhance product quality.

The advancements in textile technology, as detailed by Zhang et al. [30], include processes from scutching to warp beam preparation, focusing on eliminating impurities and defects to enhance production quality and efficiency. Moreover, the industry's need for modernization and equipment upgrades underscores the importance of updating machinery to meet contemporary standards and boost competitiveness. Overall, mechanization in textile production is crucial for fostering progress, increasing productivity, and ensuring high-quality textile products.

3.1.2 Advancement of Quality Control Approaches

Automation has flagged the way for superior quality standards through lessening errors. Machines are generally more precise and less prone to mistakes than humans. Unlike humans, automated equipment doesn't suffer from fatigue, allowing it to perform demanding tasks consistently without needing breaks. With advancements in technology, human errors have significantly decreased, resulting in higher-quality products at lower costs, making them more affordable for consumers [31].

3.1.3 Development of Better Fabrics

Technology plays a crucial role in producing high-quality fabrics by driving various advancements in the textile industry. Additive manufacturing technology, as discussed by Dehghani and Goyal [32], enables cleaner textile production through 3D printing, reducing material waste and energy consumption. Digital imaging and learning methods aid in flaw detection, enhancing product quality by identifying and rectifying fabric defects. Fabric manufacturing technologies emphasize sustainability by reducing energy consumption and environmental impact. Additionally, textile technologies streamline production processes, improve textile quality, and increase efficiency by removing impurities, ensuring uniformity, and enhancing fabric processing quality. Overall, technology plays a multifaceted role in elevating fabric quality through innovation, sustainability, and efficiency improvements [33].

3.1.4 Utilization of Computers in Design Development

Despite the use of computers in textile design, the industry has undergone a revolution as a result of increased efficiency, automation, and product innovation. Computer applications have facilitated the transition from conventional to technical textiles, enabling close observation of the whole manufacturing process, from the fibre to the finished garment or any textile product, and adapting to changing market dynamics for competitive survival [34].

In the textile industry, computer-aided design (CAD) software is essential for quickly developing new designs, optimizing production procedures, and facilitating the development of complex weave patterns and combinations for a broad spectrum of textile merchandise. Furthermore, the incorporation of artificial intelligence into digital textiles has opened up opportunities for intelligent and user-friendly wearable experiences, emphasizing the significance of tactile interaction and distinct fabric properties in the creation of interfaces [35]. Additionally, the use of user-friendly apps like Virtual Prototyping tools in the integration of electronic textiles into fashion education is improving students' programming and hands-on computer skills. Future applications in fashion colleges and allied design sectors are being paved by this progress [36].

3.2 Investigating Fascinating Novel Technologies

Innovations in textile design are driven by technology, and these developments will continue to grow in the years to come. Technology empowers designers to create innovative concepts and turn them into real-life possibilities, from basic clothing to luxurious outfits that bring dreams to life [29]. In the future, we may have garments that adapt to both summer and winter conditions, featuring digital sensors that adjust their texture based on the external environment. With the advancements in fabric design capable of enduring the harsh elements of outer space, the possibilities seem limitless.

The field of textile technology has advanced to amazing heights, and its applications will continue to get better in the years to come. Textiles and fashion are inextricably linked, therefore there will be many more fascinating trends to come. The rapid advancement of technology has accelerated the quest for better materials, distinctive styles, and more functional clothing that will enhance everyday living [37].

The textile industry has experienced a technological revolution that has transformed the processes involved in creating fabrics. This has led to the development of smart textiles that possess unique features including flame retardancy, high flexibility, and thermal insulation [25]. These developments have also sparked a shift toward digital transformation, with investments in robots, automation, cloud computing, and machine learning (ML) technologies greatly boosting the sustainability and competitiveness of businesses. Furthermore, efforts to use eco-friendly chemicals and reagents to lower energy consumption and environmental pollution in textile manufacturing have changed the focus to sustainable development [38]. In addition, in order to overcome the environmental problems caused by traditional water dyeing methods, novel and cutting-edge sustainable dyeing technologies like air dyeing, supercritical carbon dioxide (Sc-CO_2) dyeing, and plasma-assisted dyeing are being investigated [39].

4 Three-Dimensional Printing in Textile Design

The technique of three-dimensional printing (3DP) has become increasingly popular in the last few decades because of its direct fabrication method, which is made possible by digital design. The inception of 3DP technology can be traced back to

Charles Hull's 1983 creation of stereolithography (SLA), which established the groundwork for modern additive manufacturing technologies. Subsequent advancements in the 1980s and 1990s, like *fused deposition modelling* (FDM) and *selective laser sintering* (SLS), broadened its applications across various industries, ultimately making 3D printing a vital component of contemporary manufacturing and design [40].

3D textile printing has transformed the textile sector, allowing designers to surpass the limitations of conventional textile manufacturing. The inception of three-dimensional printing (3DP) dates back to the early 2000s, when designers and researchers started incorporating additive manufacturing techniques [41]. In the twenty-first century, apparel designers and textile manufacturers are leveraging increasingly affordable and advanced 3D-printing technology to reduce the environmental impact of fast fashion. This innovation promotes sustainability by enabling the recycling of existing materials and lowering production and resource consumption. Today's fashion designers and textile manufacturers are boosting sustainability by recycling pre- and post-consumer waste, cutting down on production waste, and minimizing the effect of fast fashion on the environment by utilizing more accessible and sophisticated 3D printing technology [42].

3D printing, also known as additive manufacturing (AM), is a method of creating 3D objects by layering materials together without the use of design templates. Unlike subtractive manufacturing technologies, 3DP shapes objects without supplementary apparatus or material waste, as it adds material layer by layer. Therefore, this technology is considered sustainable and eco-friendly [40, 43]. In essence, 3D printing forms three-dimensional objects using printing material applied through custom-made or standard printing heads. The hot end, extruder, print bed, nozzle, cooling fan, and build area are the key parts of a 3D printer (Fig. 1) [44, 45].

In recent decades, several methods for 3D printing have been invented, including digital light processing (DLP), selective laser sintering (SLS), fused filament fabrication (FFF), selective laser melting (SLM), stereolithography (SLA), and fused deposition modelling (FDM) [41].

Compared with conventional production methods, 3DP techniques have many benefits, including reduced material waste, higher design freedom, and cost, time, resource, and energy savings. Numerous industries have thoroughly investigated and extensively employed these 3D printing technologies, such as manufacturing, space exploration, transportation, aerospace, and construction [40, 41].

4.1 *Types of Three-Dimensional Printing*

Seven main categories of 3D printing procedures are recognized by ASTM Standard F2792 [46], as illustrated in Fig. 2: Binder jetting, material extrusion, material jetting, powder bed fusion, sheet lamination, and vat photopolymerization.

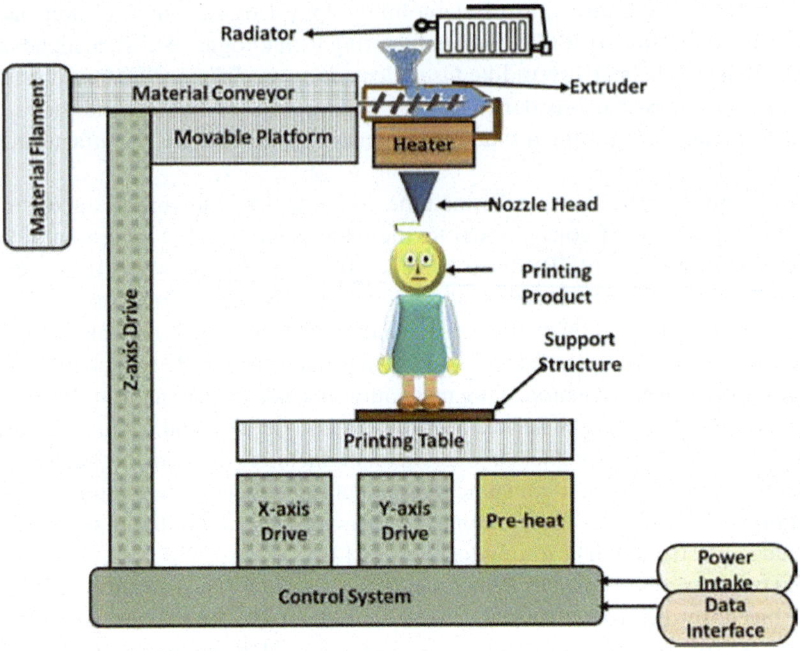

Fig. 1 Schematic diagram of a 3D printing device [44]

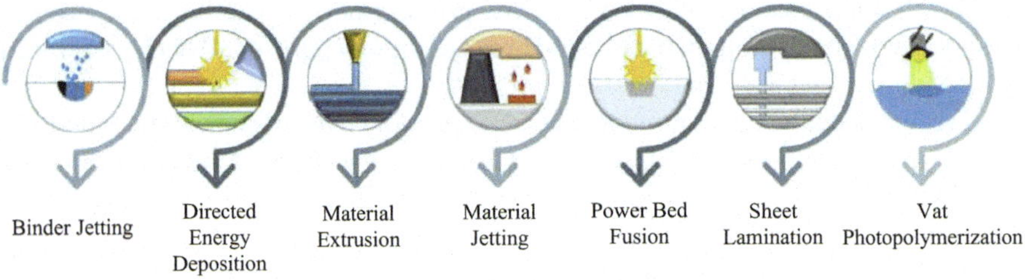

Fig. 2 Seven principal classifications of 3D printing techniques (ASTM Standard F2792) [40]

4.1.1 Vat Photopolymerization

In this, a model is created by layering using a vat containing liquid photopolymer resin. The resin is selectively thickened and hardened via light-activated polymerization, usually using a UV laser or a digital light processing (DLP) projector [47].

4.1.2 Material Jetting

Material jetting dispenses build material droplets, which are cured by UV light to develop into layers and are used to manufacture objects. This technique achieves high dimensional precision and a smooth surface appearance, and it also supports multi-material and multi-colour printing [48].

4.1.3 Binder Jetting

Powder materials are bound together via binder jetting, which employs a liquid bonding agent. A print head moves horizontally and vertically, depositing the adhesive onto a powder bed to fabricate parts layer by layer. It accommodates a broad range of raw materials, such as sand, metals, ceramics, and polymers [47].

4.1.4 Material Extrusion

Material extrusion, alternately known as fused deposition modelling (FDM) or fused filament fabrication (FFF), utilizes a nozzle to extrude thermoplastic or composite material in a semi-molten state onto a build platform. The material is layered to create the finished product. Among other 3DP techniques, this is one of the most popular methods available for printing [46, 49].

4.1.5 Powder Bed Fusion

In this technique the particular region of the powder bed is fused together using thermal energy. This category includes a variety of techniques such as selective laser melting (SLM), electron beam melting (EBM), and direct metal laser sintering (DMLS). The powder particles are heated and fused together using a high-energy source, such as a laser or electron beam. [50].

4.1.6 Sheet Lamination

In this method the layers of material are fused together to create a product. This includes methods like laminated object manufacturing (LOM) and ultrasonic additive manufacturing (UAM). Specifically, UAM combines layers of metal sheets by employing ultrasonic sound waves [51].

4.1.7 Directed Energy Deposition (DED)

This process involves the use of a nozzle that may move in several axes to deposit molten materials precisely where needed on a substrate. This method can be applied to both creating new parts from scratch and adding material to already existing ones [52].

4.2 *Software Tools for 3D Printing of Textiles*

Software tools for 3D printing in textiles are essential for designing, simulating, and optimizing textile structures and patterns. These tools facilitate the creation of complex designs and ensure efficient and accurate production. Some of the key software used in 3D modelling and design, slicing and printing, simulation and visualization, and data management is shown in Fig. 3 [42, 53–56].

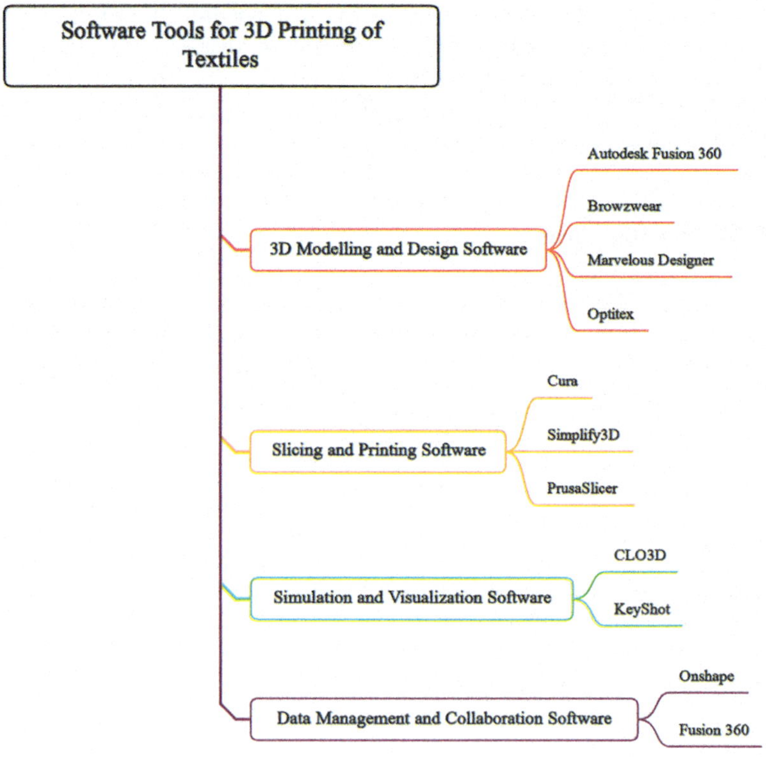

Fig. 3 Software tools for 3D printing of textiles [42, 53–56]

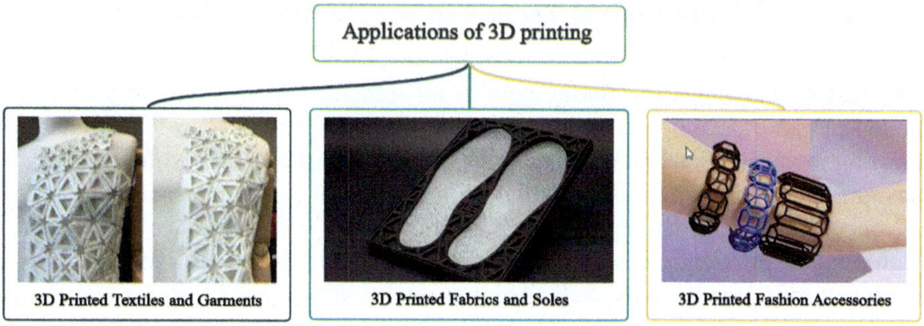

Fig. 4 Applications of 3D printing in textiles (Image availed from [57]—3DP Garment, [58]—Sole and [59]—Fashion accessories)

4.3 Utilizing 3D Printing in Textile Manufacturing

There are several uses for three-dimensional printing in the textile industry. The key areas of 3DP in textiles are depicted in Fig. 4.

4.3.1 Three-Dimensional Printed Textiles and Garments

3D printing technology can revolutionize textile structures and garment production by streamlining the traditional processes of cutting and sewing [55, 60, 61]. Techniques such as SLS and FDM are widely utilized to create textile materials [42, 61]. These methods improve the production of intricate designs and patterns with high accuracy and efficiency. Popular materials like thermoplastic polyurethane (TPU) are favoured for their plasticity, endurance, and ability to replicate traditional textile properties [42, 62].

4.3.2 Three-Dimensional Printed Fabrics and Soles

Though 3D printing innovative fabric patterns and structures can be fabricated while it is challenging by means of conventional methods. Major brands, such as Nike, have adopted 3D printing to develop customized, high-performance footwear, including soles and uppers. Additionally, 3D-printed metal fabrics can be manufactured, offering benefits like reduced weight for specialized applications [61].

4.3.3 Three-Dimensional Printed Fashion Accessories

Designers leverage 3D printing to craft intricate and customized fashion accessories, such as jewellery, headpieces, and other decorative elements. The design freedom provided by 3D printing facilitates the exploration of complex geometric shapes and patterns in the creation of fashion accessories [42, 55].

5 Future Technological Trends in the Textile Sector

The textile sector has undergone significant transformation in recent years, propelled by rapid developments in various technologies. From smart textiles and 3D printing to automation and data analytics, these innovations are revolutionizing the design, production, and consumption of fabrics and garments. Technological advancements in weaving, dyeing, finishing, and printing have significantly increased the efficiency, environmental friendliness, and cost-effectiveness of fabric manufacturing. Innovations such as higher-quality fibres and AI-driven textiles are revolutionizing our perception and production of fabrics, making the industry more dynamic than ever. Numerous factors, from the advancement of novel materials and technologies to the shifting global marketplace, are influencing how textiles are produced and sold [63]. Here are a few ways technology is transforming the industry:

- **3D technology**: Speeds up the production of textile goods, making the process faster than before.
- **Automation and robotics**: Enhance manufacturing efficiency and precision.
- **Sustainable manufacturing**: Introduces more environmentally friendly processes.
- **Consumer demand for sustainability**: Drives the industry towards producing more sustainable products.

5.1 *Novel Textiles*

Advancements and innovative technologies in materials science are increasingly focused on adopting ethical practices. To lessen environmental impact, producers are integrating new materials that enhance traditional functionalities. Innovations like plant-based textiles, hydrophobic cotton, antimicrobial fabrics, as well as shape memory polymers exemplify the adaptability of modern textile advancements. These developments provide environmentally friendly substitutes that can be produced on a big scale and are profitable [64].

5.1.1 Recycled Fabric by *Refiberd*

US-based start-up Refiberd uses only post-consumer waste to make its recycled textile products. Refiberd uses robotics and artificial intelligence (AI) to turn old clothes into new threads for apparel using its patent-pending technology. Using an environmentally benign chemical procedure, the start-up separates mixed garbage and turns it into textile threads made of cellulose and polyester. This strategy enables consumers and clients to embrace reasonably priced and environmentally friendly textile substitutes [65].

5.1.2 Reversible Textile Colouring by *Vividye*

Swedish start-up *Vividye* has developed a reversible textile colouring technology that uses resource-efficient printing methods. This innovative technology allows for the multiple decolouring and recolouring of old fabrics. *Vividye's* chemical formulations gently remove designs without damaging the material, thereby increasing the lifecycle of fabrics. By doing so, Vividye standardises water resource usage and reduces the number of chemicals released into the environment [66].

5.2 Advanced Apparel Manufacturing

Advanced apparel manufacturing technology is evolving to prioritize energy efficiency and high-speed processing. Garment manufacturing traditionally demands substantial resources, including materials, chemicals, water, and energy, which results in the creation of material waste and causing environmental pollution. ICT-based systems, CAD software, rapid digital printing, and robotics are some of the most recent innovations. By increasing automation and precision in production, quality control, and human resource management, these technologies raise overall sustainability and efficiency [64].

5.2.1 Automation and Robotics

Automation has revolutionized the textile and garment manufacturing industry, permeating every aspect, from multifunctional sewing machines to robotic arms handling repetitive tasks. The primary objectives are to enhance efficiency, reduce labour costs, and improve product quality [4, 67].

- **Robotic sewing**: Machines now manage the intricate task of stitching various fabrics with exceptional precision and speed, surpassing human capabilities. These machines operate tirelessly and consistently.
- **Automated cutting**: Technological advancements have significantly improved the cutting process, offered superior precision and minimized material waste. This resulted in cost savings and accelerated production times.
- **AI-run factories**: AI-driven automation reduces the need for human labour, leading to lower labour costs and increased productivity. This advancement could potentially lead to reshoring, reducing reliance on overseas production facilities and enhancing competitiveness.

5.2.2 Digital Manufacturing Management by *Coats Digital*

US-based start-up *Coats Digital* uses tech-driven solutions to improve digital industrial management. The company uses AI and Big Data Analytics to automate important production processes including fabric sourcing, design development, and shop floor executions. Their VisionPLM software provides precise estimations of operating expenses, raw material capacity, and fabric consumption. As a result, *Coats Digital*'s software solutions simplify the administration of crucial processes, rendering them economically viable and long-lasting [68].

5.2.3 Intelligent Robotics by *Sewts*

Sewts is a German start-up which integrates robotics into apparel and textile manufacturing by leveraging computer vision technology and machine learning (ML) tools. Their image-processing software facilitates the management of textiles and materials that are easily malleable. For instance, the Velum System automatically folds clothes to speed up washing procedures. This is made possible by artificial intelligence systems that provide real-time behaviour predictions for dimensionally unstable materials. The technically difficult task of robotic textile handling is made possible by *Sewts*, which greatly speeds up the production of apparel [69].

5.3 *Advanced Three-Dimensional Technologies*

The advent of 3DP innovations is significantly boosting the efficiency of textile design as well as manufacturing. 3D printing makes it easier to experiment with novel shapes, structures, and prototypes. This permits the production of a wide range of materials, including soft polymers and stiff ceramics. This technology helps zero-waste initiatives by boosting the use of sustainable materials. Additionally, 3D design software enables the virtual visualization of garments, reducing fabric waste and unnecessary production [64].

5.3.1 Three-Dimensional (3D) Knitwear by *New Industrial Order*

Dutch start-up *New Industrial Order* is tackling fast fashion waste by manufacturing 3D-printed knitwear. Their 3D machines produce garments in one piece, eliminating the need for cutting and sewing. This method also allows the yarn to be unravelled and reused for knitting. Furthermore, the start-up's Knitcloud platform digitizes the 3D knitwear supply chain and enhances the circular economy within the fashion sector when paired with digital production [70].

5.3.2 Three-Dimensional (3D) Sports Equipment by *Shapeshift*

Canadian start-up Shapeshift is a 3D printing company that uses artificial intelligence (AI) to create personalized sports and medical equipment. Their platform uses body scanning technology to customize items like bicycle helmets and prosthetic limbs. These 3D scans are analysed by intelligent algorithms which optimize fit, form, and function. 3D printers use this data to quickly and extensively make products. The mass creation of personalized wearables with bespoke fits is made more accessible by *Shapeshift*'s 3D technology [71].

5.4 Sustainability in Textile Supply Chain

As the market for affordable apparel grows, companies have set up supply chains to create large quantities of synthetic clothes made from petroleum derivatives. In response, technologically advanced tactics are accelerating the fast fashion industry's shift to closed-loop systems and a circular economy. Reuse, recycling, and upcycling are eco-friendly activities that start-ups are pushing to accelerate this change [64, 72].

5.4.1 Digitalization of Fashion Supply Chain by *eRMG*

Singaporean start-up *eRMG* offers a SaaS platform for brands, wholesalers, and retailers to streamline post-pandemic garment sourcing. Real-time tracking, virtual order administration, data-driven procedures, and a digital design studio are some of their digital technologies. This technologically advanced solution enhances supplier validation, ensuring reliability and compliance [64].

5.4.2 Decentralized Supply Chain Platform Built by *Digitalax*

US-based start-up *Digitalax* has created a decentralized platform for fashion supply chains to track and authenticate digital goods by employing Ethereum and NFTs. The programme generates new patterns and textile libraries from raw elements and converts them into digital assets with dynamic pricing. This democratizes the Web3 supply chain by creating a fashion NFT auction exchange that is exclusively digital [73].

5.5 *Artificial Intelligence and Data Analytics in the Textile Sector*

AI and data analytics improve management automation, product inspection, and business optimization. These innovations sense fabric defects, evaluate wrinkles, uncover hidden operating patterns, and track consumer behaviour for better recommendations. This leads to improved workflows, labour management, and product quality [64].

5.5.1 Tote Delivers AI Apparel Tagging

Spanish start-up *SENSTILE* uses sensors and AI to digitize textile substrate properties, analysing visual and chemical characteristics, and interpreting fabric feel and touch. Their platform, *FabrikHUB*, converts this data into digital identification files, helping to find suitable suppliers and reduce fashion waste [74].

5.5.2 AI-Based Design Automation by *Yoona.ai*

Yoona.ai is a German start-up that offers AI-driven design process automation tools. Using machine learning, it analyses fashion input data like drawings, photos, and prototypes to generate digital designs. The platform suggests new collections, prints, materials, and colour schemes, optimizing design workflows to be more time-efficient, cost-saving, and environmentally friendly [75].

5.6 *Augmented Reality and Virtual Reality*

The inception of augmented reality (AR) and virtual reality (VR) are revolutionizing the textile and apparel industry. AR enables consumers to see how garments will look and fit in real-time through virtual try-ons, significantly enhancing the shopping experience. VR provides immersive environments for designers and manufacturers to collaborate, prototype, and test products virtually, minimizing the need for physical samples. These technologies streamline the design process, boost customer engagement, and help reduce costs and waste [76]. Retailers and producers are utilizing AR and VR in various ways within the fashion segment, including virtual try-ons, virtual fitting rooms, interactive display screens, virtual makeovers, virtual tailors, and in-store navigation, shown in Fig. 5. Renowned brands such as Zara, Ralph Lauren, H&M, Converse, Louis Vuitton, Gucci, Burberry, and Chanel are utilizing AR and VR to market a wide range of clothing, shoes, and accessories [77].

Fig. 5 Transformative uses of AR & VR in the fashion sector

6 Conclusion

The textile industry has undergone a substantial transformation in the years due to rapid technological advancements. The integration of CAD software, 3DP technology, e-textiles, and digital textile printing has revolutionized textile design, production, and product development. CAD tools have transformed how designers create textile and apparel designs, while 3D printing has introduced new possibilities for crafting complex, customized textile structures and garments. Smart textiles now incorporate sensors, electronics, and interactive features, enhancing their functionality, and digital textile printing allows for the direct application of digital designs onto fabrics, enabling unique and personalized textile patterns without traditional printing limitations. Additionally, the integration of data analytics and artificial intelligence (AI) into the design process has optimized the industry's capability to meet market demands and consumer preferences. These advancements in technology have aided the creation of more innovative, personalized, and sustainable textile products, and they will continue to reform the future of textile design as the industry evolves.

References

1. Anonymous. (n.d.). *Technological developments in textiles*. Retrieved July 9, 2024, from https://courses.lumenlearning.com/suny-hccc-worldhistory2/chapter/technological-developments-in-textiles/

2. Audaces. (2021). *Textile technology: 4 innovations that will reshape fashion*. https://audaces.com/en/blog/textile-technology
3. Shah, D. (2023). *Impact of digital technologies on the textile industry*. Fibre2Fashion. https://www.fibre2fashion.com/industry-article/9622/impact-of-digital-technologies-on-the-textile-industry
4. Amasa, Y. (2023). *The future of textile and garment manufacturing: Top 10 tech trends*. Syndell. https://syndelltech.com/top-10-tech-trends-for-the-textile-and-garment-manufacturing-industry/
5. Browzwear.com. (n.d.). *Browzwear: 3D fashion design software*. Retrieved July 9, 2024, from https://www.browzwear.com/
6. Optitex.com. (n.d.). *Optitex–Fashion design and production software*. Retrieved July 9, 2024, from https://www.optitex.com/
7. Lipson, H., & Kurman, M. (2013). *Fabricated: The new world of 3D printing*. John Wiley & Sons.
8. Rissanen, T., & McQuillan, H. (2023). *Zero waste fashion design*. Bloomsbury.
9. Cherenack, K., & Van Pieterson, L. (2012). Smart textiles: Challenges and opportunities. *Journal of Applied Physics, 112*(9), 091301. https://doi.org/10.1063/1.4742728
10. Stoppa, M., & Chiolerio, A. (2014). Wearable electronics and smart textiles: A critical review. *Sensors, 14*(7), 11957–11992. https://doi.org/10.3390/s140711957
11. Dunne, L. E., Ashdown, S. P., & Smyth, B. (2005). Expanding garment functionality through embedded electronic technology. *Journal of Textile and Apparel Technology and Management, 4*(3), 1–11.
12. Baurley, S. (2004). Interactive and experiential design in smart textile products and applications. *Personal and Ubiquitous Computing, 8*, 274–281. https://doi.org/10.1007/s00779-004-0288-5
13. Ujiie, H. (2006). *Digital printing of textiles*. Elsevier. https://www.sciencedirect.com/book/9781855739512/digital-printing-of-textiles
14. Ujiie, H. (2014). *Digital textile design*. Laurence King Publishing.
15. Choi, T. M., Cai, Y. J., & Shen, B. (2018). Sustainable fashion supply chain management: A system of systems analysis. *IEEE Transactions on Engineering Management, 66*(4), 730–745. https://doi.org/10.1109/TEM.2018.2857831
16. Niinimäki, K. (2013). *Sustainable fashion: New approaches*. Aalto University. https://www.academia.edu/68551167/Sustainable_Fashion_New_Approaches
17. Gwilt, A. (2020). *A practical guide to sustainable fashion*. Bloomsbury. https://www.bloomsbury.com/in/practical-guide-to-sustainable-fashion-9781350067042/
18. Wang, H., & Memon, H. (2023). Introduction to digital textile printing. In *Digital textile printing: Science, technology and markets* (pp. 1–19). Elsevier. https://doi.org/10.1016/b978-0-443-15414-0.00001-7
19. Kumelachew, D. M., Wagaye, B. T., & Adamu, B. F. (2023). Digital textile printing innovations and the future. In *Digital textile printing: Science, technology and markets* (pp. 241–259). Elsevier. https://doi.org/10.1016/b978-0-443-15414-0.00006-6
20. Wagaye, B. T., Kumelachew, D. M., & Adamu, B. F. (2023). Digital image design and creation of printed images on textile fabrics. In *Digital textile printing: Science, technology and markets* (pp. 73–91). Elsevier. https://doi.org/10.1016/b978-0-443-15414-0.00011-x
21. Rohan. (2023, November 19). *Modern advances in textile technology: The fabric revolution*. Knowing Fabric. https://knowingfabric.com/modern-advances-in-textile-technology-the-fabric-revolution/
22. Thompson, M. (2023, December 27). *What is digital printing on fabric: Complete guide 2024*. https://www.printkk.com/blog/articles/what-is-digital-printing-on-fabric
23. Hassan, H., & Anshun, N. (2022). Benefits of digital printing for fashion entrepreneurs: A case study at Alia Bastamam. In *International Visualization, Informatics and Technology Conference*. IEEE. https://doi.org/10.1109/ivit55443.2022.10033381

24. Babar, A. A., Gianchandani, P. K., & Jhatial, A. K. (2023). Overview of different digital textile printing machines. In *Digital textile printing: Science, technology and markets* (pp. 41–55). Elsevier. https://doi.org/10.1016/b978-0-443-15414-0.00014-5
25. Kumar, B. (2023). Advanced technology in fabric manufacturing. In *Textile science and clothing technology* (pp. 65–95). Springer. https://doi.org/10.1007/978-981-99-2142-3_3
26. Christiane, S. (2022). *Research on the accumulation effect model of technological innovation in textile industry based on chaos theory*. https://doi.org/10.48550/arxiv.2204.08340
27. Duan, Y., Zhang, M., Zhou, J., Zhou, H., Xu, L., & Yang, H. (2022). Exploring the law of color presentation of double-sided heterochromatic digital printing for textile. *Frontiers in Psychology, 13*, 956748. https://doi.org/10.3389/fpsyg.2022.956748
28. Ranga, S., Meda, U. S., & Kiran, S. (2022). Nanotechnology in the textile industry: Present and future. *ECS Transactions, 107*(1), 4791–4798. https://doi.org/10.1149/10701.4791ecst
29. Anonymous. (2019, August 8). *The role of technology in the textile industry*. Textile Consult. https://www.textileconsult.co.uk/2019/08/08/the-role-of-technology-in-the-textile-industry/
30. Zhang, L., Leung, M. Y., Boriskina, S., & Tao, X. (2022). Advancing life cycle sustainability of textiles through technological innovations. *Nature Sustainability, 6*(3), 243–253. https://doi.org/10.1038/s41893-022-01004-5
31. Patil, G., & Deshmukh, S. M. (2022). An approach to fabric defect detection using statistical methods for feature extraction. *Journal of Physics Conference Series, 2327*(1), 012033. https://doi.org/10.1088/1742-6596/2327/1/012033
32. Dehghani, M., & Goyal, P. (2022). Design and development of textile fabrics using 3D printing technology. *ECS Transactions, 107*(1), 19313–19323. https://doi.org/10.1149/10701.19313ecst
33. Rahimunnisa, K. (2022). Textile fabric defect detection. *Journal of Innovative Image Processing, 4*(3), 165–172. https://doi.org/10.36548/jiip.2022.3.004
34. Saini, N., Yadav, S., & Rose, M. N. (2018). Fabric designing for product development by combination of weaves through CAD. *Journal of Textile Science and Engineering, 8*(2), 343. https://doi.org/10.4172/2165-8064.1000343
35. Tan, J., Shao, L., Lam, N. Y. K., Toomey, A., & Ge, L. (2021). Intelligent textiles: Designing a gesture-controlled illuminated textile based on computer vision. *Textile Research Journal, 92*(17–18), 3034–3048. https://doi.org/10.1177/00405175211034245
36. Han, A., Wohn, K., & Ahn, J. (2020). Towards new fashion design education: Learning virtual prototyping using E-textiles. *International Journal of Technology and Design Education, 31*(2), 379–400. https://doi.org/10.1007/s10798-019-09558-w
37. Arikan, C. O. (2019). Developing an intelligent automation and reporting system for fabric inspection machines. *Tekstil Ve Konfeksiyon Dergisi, 29*(1), 93–100. https://doi.org/10.32710/tekstilvekonfeksiyon.491362
38. Khalil, E., Sarkar, J., Rahman, M. M., Shamsuzzaman, M., & Das, D. (2023). Advanced technology in textile dyeing. In *Advanced technology in textiles: Fibre to apparel* (pp. 97–138). Springer Nature. https://doi.org/10.1007/978-981-99-2142-3_4
39. Fazal, M. Z., & Hamdani, S. T. A. (2023). 12 computer applications in textiles. In *Textile engineering* (pp. 217–228). De Gruyter. https://doi.org/10.1515/9783110799415-012
40. Dip, T. M., Emu, A. S., Nafiz, M. N. H., Kundu, P., Rakhi, H. R., Sayam, A., et al. (2020). 3D printing technology for textiles and fashion. *Textile Progress, 52*(4), 167–260. https://doi.org/10.1080/00405167.2021.1978223
41. Davey, R. (2022). *How is 3D printing changing the textile industry?* AZO Materials. https://www.azom.com/article.aspx?ArticleID=21953
42. Barnes, J. (2023). *3D printing for textiles*. AATCC Newsletter. https://www.aatcc.org/aatccnews_06a/
43. Campbell, T., Williams, C., Ivanova, O., & Garrett, B. (2021). *Could 3D printing change the world?* http://www.atlanticcouncil.org/publications/reports/could-3d-printing-change-the-world

44. Peng, T. (2016). Analysis of energy utilization in 3D printing processes. *Procedia CIRP, 40*, 62–67. https://doi.org/10.1016/j.procir.2016.01.055
45. Motadaka, S. (2019). *A review of 3D printing research from computer and information science perspective: Challenges and research directions*. University at Buffalo.
46. ASTM F2792 Standards. (n.d.). *7 families of additive manufacturing*. https://www.additive-manufacturing.media/cdn/cms/7_families_print_version.pdf
47. Low, Z. X., Chua, Y. T., Ray, B. M., Mattia, D., Metcalfe, I. S., & Patterson, D. A. (2017). Perspective on 3D printing of separation membranes and comparison to related unconventional fabrication techniques. *Journal of Membrane Science, 523*, 596–613.
48. Yap, Y. L., Wang, C., Sing, S. L., Dikshit, V., Yeong, W. Y., & Wei, J. (2017). Material jetting additive manufacturing: An experimental study using designed metrological benchmarks. *Precision Engineering, 50*, 275–285.
49. Kaynak, B., Spoerk, M., Shirole, A., Ziegler, W., & Sapkota, J. (2018). Polypropylene/cellulose composites for material extrusion additive manufacturing. *Macromolecular Materials and Engineering, 303*(5), 1800037–1800038.
50. Ventola, L. C. (2014). Medical applications for 3D printing: Current and projected uses. *P T, 39*(10), 704–711.
51. Chang, B., Parandoush, P., Li, X., Ruan, S., Shen, C., Behnagh, R. A., Liu, Y., & Lin, D. (2020). Ultrafast printing of continuous fiber-reinforced thermoplastic composites with ultrahigh mechanical performance by ultrasonic-assisted laminated object manufacturing. *Polymer Composites, 41*(11), 4706–4715.
52. Gibson, I., Rosen, D., & Stucker, B. (2010). Introduction and basic principles. In *Additive manufacturing technologies* (pp. 20–35). Springer. https://doi.org/10.1007/978-1-4419-1120-9_1
53. Wu, S., Zeng, T., Liu, Z., Ma, G., Xiong, Z., Zuo, L., & Zhou, Z. (2022). 3D printing technology for smart clothing: A topic review. *Materials, 15*(20), 7391. https://doi.org/10.3390/ma15207391
54. Xiao, Y. Q., & Kan, C. W. (2022). Review on development and application of 3D-printing technology in textile and fashion design. *Coatings, 12*(2), 267. https://doi.org/10.3390/coatings12020267
55. Manaia, J. P., Cerejo, F., & Duarte, J. (2023). Revolutionising textile manufacturing: A comprehensive review on 3D and 4D printing technologies. *Fashion and Textiles, 10*(1), 20. https://doi.org/10.1186/s40691-023-00339-7
56. Kuska, A. (2023). *What is the best clothing design software for 3D rendering & modeling apparel?* Cad Crowd. https://www.cadcrowd.com/blog/what-is-the-best-clothing-design-software-for-3d-rendering-modeling-apparel/
57. Kang, M., & Kim, S. (2019). Fabrication of 3D printed garments using flat patterns and motifs. *International Journal of Clothing Science and Technology, 31*(5), 653–662. https://doi.org/10.1108/IJCST-02-2019-0019
58. Carlota, V. (2019). *Customised and 3D printed insoles from ECCO*. 3D Natives. https://www.3dnatives.com/en/ecco-printed-insoles-010320194/
59. Upputuri, R. S. (2016). *3D printing in consumer products*. THINK3D. https://www.think3d.in/3d-printing-in-consumer-products-industry/
60. Chatterjee, K., & Ghosh, T. K. (2020). 3D printing of textiles: Potential roadmap to printing with fibers. *Advanced Materials, 32*(4), 1902086. https://doi.org/10.1002/adma.201902086
61. IN3DTEC. (2023). *The application of 3D printing in fashion*. https://www.in3dtec.com/the-application-of-3d-printing-in-fashion/
62. Čuk, M., Bizjak, M., Muck, D., & Kočevar, T. N. (2020, November). 3D printing and functionalization of textiles. In *10th International Symposium on Graphic Engineering and Design* (pp. 499–506). https://doi.org/10.24867/GRID-2020-p56
63. FibreGuard. (2022). *How technology is changing the textile industry*. https://fibreguard.com/sell/blog/how-technology-is-changing-the-textile-industry

64. StartUs. (2024). *Discover the top 8 trends in the textile industry (2024)*. https://www.startus-insights.com/innovators-guide/textile-industry-trends/#:~:text=5.-,3D%20Technologies,soft%20polymers%20to%20rigid%20ceramics
65. Refiberd. (n.d.). *Intelligent sorting for textile-to-textile recycling*. https://refiberd.com/
66. Vividye. (n.d.). *High performing sustainable coloring*. https://www.vividye.com/
67. Somani, V. (2023). *10 Technology trends defining the future of textile and garment manufacturing*. https://www.worldfashionexchange.com/blog/tech-trends-of-textile-and-garment-manufacturing/
68. Coats Digital. (n.d.). *Leading the digital transformation of the fashion industry*. https://www.coatsdigital.com/en/manufacturer/
69. Sewts. (n.d.). *Sewts—Automated handling with physical AI*. https://www.sewts.com/home
70. NIO. (n.d.). *New Industrial Order*. https://new-industrial-order.com/
71. Shapeshift. (n.d.). *Shapeshift 3D*. https://www.shapeshift3d.com/custom-fit-software
72. Sampson, L. (2023). *Fashion supply chain: Everything you need to know*. Oracle. https://www.oracle.com/in/retail/fashion/fashion-supply-chain/
73. Digitalax. (n.d.). https://www.digitalax.xyz/
74. Senstile. (n.d.). *AI and textiles: A touching collaboration*. https://www.senstile.com/
75. Yoona AI. (n.d.). *Yoona AI-the best AI design tool, fine-tuned for the fashion industry*. https://www.yoona.ai/what-we-do
76. Kiron, M.I. (2021). *Augmented reality (AR) in fashion industry*. Textile Learner. https://textilelearner.net/augmented-reality-in-fashion-industry/
77. Textile Today. (2022). *Augmented reality (AR) and virtual reality (VR) in fashion industry*. https://www.textiletoday.com.bd/augmented-reality-ar-virtual-reality-vr-fashion-industry

Inspired from Nature

Anthima Ram

1 Introduction

Throughout Asia textiles are one of the most powerful and exciting art forms, and in Southeast Asia in particular, the spiritual and ritual importance that textiles play in ceremonies of state and religion is reflected in their great mystery and splendour. Southeast Asian textiles are outstanding works of art, formed by a rich variety of techniques. The finest examples, often of elaborate and complex design, display superb levels of technical skill in weaving, dyeing, embroidery and appliqué. A diversity of materials includes bark, plant fibres, cotton, silk, beads, shells, gold and silver, and among a profusion of patterns and motifs, we find human figures, abstract geometric shapes, ships, arabesques, calligraphy, flowers, recognizable animals and imaginary monsters. The most common function of textiles is their use as articles of clothing. However, apart from their importance as everyday and ceremonial dress, textiles in Southeast Asia have numerous other functions including their use as religious hangings, royal insignia, theatrical backdrops, sacred talismans or secular currency, for they are intimately connected to systems of religion, political organization, marriage, social status and exchange. These functions, in turn, affect the size, shape, structure and decoration of the cloths [1].

As ornamental textiles are valuable as ornate costumes for celebrations, status symbols, objects of wealth, and religious objects, producing them frequently necessitates taking material and spiritual safety precautions in order to preserve the quality of the weaving and dying as well as the health of the craftsperson. As a result, there are myths and customs surrounding the creation of valuable textiles as well as

A. Ram (✉)
Department of Fashion Design, MIT Institute of Design, MIT ADT University,
Pune, Maharashtra, India
e-mail: anthima.ram@mituniversity.edu.in

Sadhna et al. (eds.), *The Art and Craft of Modern Textile Design*, Sustainable
Textiles: Production, Processing, Manufacturing & Chemistry,
https://doi.org/10.1007/978-3-031-86797-2_5

their origins. The standard criteria for evaluating the beauty and worth of these textiles are the texture of the materials, the artistry of the craftswoman, the depth of colour, and the clarity and complexity of the patterning and design. But, as we'll see, certain plain-dyed or striped fabrics contain a lot of ceremonial power. Furthermore, a lot of patterns and designs express vital ideas that are only understood by individuals who are familiar with the specific social and religious beliefs of the people who created them. We can only start to comprehend the actual worth and significance of clothing by viewing it in its cultural context [1].

According to the historical record, there has always been a strong connection between nature and textile design, going back many years in the annals of art and skill. Across many countries and centuries, botanical patterns and organic themes have been a mainstay of textile design, signifying humanity's continuous affinity with the natural world.

1.1 Historical Civilizations

Textiles from ancient Egypt were decorated with floral and botanical themes, which expressed the civilization's respect for the natural world and its symbolic implications. Common design motifs were papyrus plants, palm fronds, and lotus flowers, which stood for rebirth and endless life. Oriental Persuasion Botanical prints have long been a staple of textile art in Asia, especially in China and Japan. Often featured on traditional clothing like kimonos and robes, the elaborate patterns of cherry blossoms, chrysanthemums, and bamboo encapsulated the delicate beauty of the natural world [2].

1.2 Colonial and Victorian Eras

European textile patterns were influenced by exotic flora and flowers brought back from all over the world during the colonial era. Floral patterns were prevalent throughout the Victorian era. Detailed botanical themes adorned everything from clothing to wallpaper, highlighting the rich and complex aspects of the natural world. Natural history was a subject of renewed interest throughout the European Renaissance [2]. During this time, the growing area of botanical science served as an inspiration for the intricate floral depictions found in textile designs.

Nature-inspired designs gained more popularity in the nineteenth century because of artists like William Morris, who emphasized the harmony and beauty inherent in natural shapes.

Alexander McQueen is among the most recognizable designers, and he is well-known for his use of organic themes. The late designer frequently used natural organic shapes and textures in her designs to create dramatic and powerful pieces. For example, McQueen's "Plato's Atlantis" collection from 2010 had biomorphic, flowing designs that resembled the swaying shapes of marine life. Fashion and natural beauty were combined in a way that was both inventive and firmly based in organic forms, thanks to the elaborate, wave-like patterns and the tactile, scale-like elements of the clothing [2].

From the Himalayas to deserts and dense forests, nature is all around us and presents a variety of difficult terrain. Great rivers that run across the wide plains are the source of life for millions of people. Numerous Indian flora, fauna, and birds have developed rich symbolic meanings. The lotus is depicted on the toranas at Sanchi as a stunning, full-bodied flower with strong stems that twist up the gate supports. Unquestionably a sign of purity, the lotus is one of the most well-known and often-used motifs. It may be carved or painted in a plethora of styles in contexts other than Buddhism. It even entered one of Delhi's most notable early Muslim structures, the Quwwat ul-Islam Mosque.

The peacock, which is currently the national bird of the republic, is a significant emblem for all Indian civilizations and has always stood for beauty in all its forms. Hindu mythology depicts the peacock and the turtle as carrying the globe on their backs. The rose gained significance in Muslim symbolism; poets perceived the face of the beloved in it, while Sufi sages regarded it as an emblem of heavenly majesty. The cypress tree, a motif brought over from Persia, is nearly always upright and straight, and its clouds represent the beloved's graceful, petite frame.

The pot, with its overhanging leaves and its version, the vase of flowers, represented rebirth, wealth, and well-being for Muslims. Its origins can be found in the Mesopotamian and Persian "Tree of Life." Many of the walls of Rajasthani palaces, adorned with one of the most joyful representations of the riches of nature, are covered in trees, plants, flowers, birds, and other creatures. By their very nature, motifs intended as solitary elements draw attention [3].

The study of the literature highlights the significance of motifs in a select few Indian traditional textiles. According to a recent article, a majority of the designs seen in Indian traditional textiles are drawn from familiar objects, plants, animals, and architecture. Sages and mythology, for example, refer to lucky symbols like Indian rudraksha, kalasha, and shankha (conches). Many Hindu gods and goddesses carry conch, which is blown for wealth and purity. The water pitcher known as a kalasha is used for ritualistic and daily purposes. It is said to represent the womb, fertility, and fresh starts. Lotus is presented during prayers and is believed to represent purity and good fortune. Buddhism accords particular value to lotuses seen in Ajanta cave paintings. With many names and meanings, the paisley pattern—also known as manga, ambi, or karrey—is utilized throughout the nation. The Mughals brought the motif to India as a fundamental component of textiles through their influence on Persian architecture and style. The term "mangai," known in Tamil Nadu, originates from Hindu mythology and is inspired by the mango fruit. Parrots have always been associated with courting and passionate love. Fish is revered because it is said to be the rebirth of Lord Vishnu, one of Hinduism's three trinity gods. The Hindu deity Indra is represented by the elephant, symbolising power, majesty, and knowledge. Both Buddhist iconography and legendary battle scenes find representation in Indian textile traditions. The rich tapestry of Indian craftsmanship is showcased through a variety of motifs, as seen in the following examples: the Benarasi sari's Konia (mango motif), Patola's depictions of tigers and elephants, Bandhej's intricate Kalasha and elephant designs, Paithani's Ashavali lotus and peacock patterns, Kanjeevaram's symbolic Rudraksha, Pashmina's Panjdar (five-petalled flower) and Chinar leaves, Kantha's recurring fish, peacock, and mango motifs, the Tsungkotepsu shawl (linked to Naga warriors), and the Mekhla's Jaapi (the Assamese headgear once worn by nobility and now utilized by farmers for protection from the elements). Each of these motifs embodies cultural heritage, spiritual significance, and regional identity [4]. The paper is an attempt to seek knowledge and understand how the natural world elements like the floral, botanical, animal, people and a combination of these become motifs, design patterns and overall textile design, giving meaning and signification to the traditional craftsmanship and traditional textiles.

2 The Journey of Indian Textiles

Indian traditional textiles evolved out of an exquisite creative drive that captured the artisans' delights, grief, challenges, and accomplishments. Textiles, out of all the handicrafts, established a class by themselves that delighted people all over the world. Rather than causing devastation, several invasions, migrations, political upheavals, and numerous other ups and downs resulted in a synthesis that improved India's textile trade. A distinctive textile heritage emerged as a result of several factors, such as regional cults, geographic variations, mineral resources, water supply, and plant and animal life. Because Indian textile crafts draw on a rich pool of stories, symbolism, imagery, and other artistic forms, they are never in a slump or stagnation. On the other hand, it grew in scope and presented an image of the Indian way of life via integration, elimination, and acceptance. Ultimately, the Indian textile industry gave rise to two distinct modes of expression: the soil-based structural treatment, which reflected the emotions and customs of the rural pastoral people bound by human bonds, and the surface treatment, which reflected the societal trend of refined court sophistication and elegance. In the case of textiles, however, the distinction between craft and art is hazy. They were categorized as crafts for a long time, but in more recent times, textile experts and art historians have given this valuable heritage the status of art. India's magnificent textiles are regarded as if they are straight out of a fairy tale because of their vivid hues, elaborate patterns that include mythology, classicism, as well as regional flare, and overall sense of beauty that has won them the moniker "exquisite poetry in colourful fabrics." These artefacts and occasional imprints of fabrics on earthenware, stoneware, and porcelain from the Harappa sites suggest that Indian textile art was already at a mature level of development, notwithstanding the inherent difficulty of researching Indian textile art in antiquity. Indian textiles reflect the people's passion for sophisticated, flowery, and embroidered apparel. Flower-patterned, painted, printed, or embroidered textiles have long been associated with clothing. Two concepts that may be summed up in two terms, ***Rupa*** (Cultivation of beauty) and ***Anuttara Jnana*** (Attainment of maximum beauty), formed the basis of Gupta art, particularly textile art. A significant number of Indian theme designs have evolved by this point. The beautiful frescoes of the Ajanta cave complex from this era may be examined attentively to identify various colour schemes and themes. The most common theme patterns are formal floral motifs or scrolls that are occasionally adorned with a lion or seared Anserine (Hamsa) motifs or bands that alternatively include geometric patterns like herringbone (rafter), orb or sphere, striations, plaid, etc. Additionally, Kalidasa observed that Indian textiles often have elegant swan motifs [5].

Early Mughal India turned to Persia for artistic inspiration, which encouraged Persian painters to leave their own country and take the style of the Safavid court with them. However, India's weaving eventually surpassed Persia in terms of unique designs and techniques. The Mughal emperors Akbar (1556–1605), Jahāngīr (1605–27), and Shah Jahān (1627–158) had a significant impact on the textile arts. Their royal ateliers produced exquisite shawls and sashes that were highly valued inside and outside India. This gorgeously constructed and designed shawl has a complex network of crushed flowers surrounded by striking black, white, and red bands. The twill tapestry weave's flattened and angular geometricized flowers expertly tile the cloth's surface, combining the design with the 25 structural components of the fabric [6].

Persian features were included as a result of Indian preferences. Human shapes and older animal and bird patterns have been replaced by floral themes. The most common embellishment in the field of Islamic art was a variety of floral themes. Figure 1 depicts a brocaded textile with metal-wrapped silk thread that depicts a

Fig. 1 A scene of two ladies striding in a field of blooming flowers accompanied by a falcon and a hound during the Safavid period (1501–1736 CE) [7]

hunting scene with two ladies walking in a field with their falcon and hound. The depicted hunting scene from the floorspread, as shown in Fig. 2, is where two men with guns are hunting the partridges. The Indian Mughal painted textile shares a very similar aesthetic perspective with the Safavid textiles. Hand-painted and hand-drawn artwork on a cotton floor spread. A buff-colored rectangle painted in vibrant hues with a predominance of red, brown, and green. A duck, a guy carrying a rifle, two men drinking together, a man toasting wine to his sweetheart, and sportsmen shooting partridges and pigeons are depicted on the long sides. These individuals are standing on a beautiful, rolling landscape with birds perched on spring trees that are yielding fruit and flowers. The entire area is covered by two sets of trees that converge at the halfway point. Between fillets of scrolling flower stems in white on a red background, the border is made up of repeating palmettes surrounded by colored floral latticework on a purple background. Between fillets of scrolling flower stems in white on a red background, the border is made up of repeating palmettes surrounded by colored floral latticework on a purple background. The panel is made up of four distinct textile pieces that have been sewn together and painted [9].

Being a naturalist at heart, Jahangir enjoyed dissecting various plants and animals. Just like him, Empress Nur Jahan was fascinated by the natural environment. During this time, a variety of visual traditions were absorbed. Indian subject painting underwent a new phase of growth when Jahangir's artists studied animals, vegetation, and flowers. A new understanding of textile theme designs was eventually brought about by Jahangir's promotion of realism, which involved an analysis of minute details that stressed distinctive qualities and realistic representation. Around the beautiful Shah Jahan period, newer designs started to develop that emphasized plants, leaves, stalks, and flowers in their natural bends and curves. The flowering plants' traits were managed with great care. Incorporating plants, leaves, sprays, blossoms in harmonies, and flowery scrolls added new aspects to studying nature.

Fig. 2 Painted and dyed cotton, Floorspread 1630 [8]

By fusing floral studies with a more distinctively Indian flavour, the Mughal art style started to take shape. Thus, the floral themes in Mughal fabrics generated new patterns for Indian textiles [9].

3 Flowers of the Garden

Flowers and flowering plants are prevalent designs in the textiles of northwestern India, Pakistan, and the Kashmir Valley, whether a professional weaver for the export market, a court artisan, or a rural lady weaving for her family. Many of the most popular flower motifs, like the Buta on Kashmir shawls, are so stylized that they may seem like simple abstract forms. This is because of the love of gardens and flowering plants that dates at least as far back as the court cultures of the early Safavid dynasty in Iran (1501–1732) and the Mughal dynasty in India (1526–1857). In the gardens of the Mughal palace, flowers blossomed and overflowed into textiles and paintings. The commissioned painting "The Flower Garden" has well depicted the flora of the garden getting converted into design motifs on textiles and art elements in paintings and architectural realm as presented in Fig. 3. Babur, the first

Fig. 3 The Flower Garden was commissioned by Emperor Jahangir in the early 1600s [10]

Mughal emperor (r. 1526–1530), planted gardens in his new realm. He provided accurate descriptions of uncommon flora, which he had court painter's paint. His compositions include poems about an apple tree in fall and spring blooms in the Hindu Kush mountains. His grandson Akbar (r. 1556–1605) used a picture of Babur directing the creation of an Iranian-style garden to illustrate his grandfather's journal, the Babur Nama [6].

This may appear to be an unexpected choice for depicting the life of a conqueror who, since leaving Ferghana at the age of 14, had spent most of his life battling. Babur's appreciation of his gardens is undoubtedly highlighted by it. Babur, his son and heir Humayun (r. 1530–1540, 1555–1556), and Akbar were art patrons and intellectuals who maintained touch with the Iranian court; Humayun was once allowed sanctuary there following a series of military setbacks in India. All three welcomed Iranian artists and craftspeople to India. Gradually, a unique Mughal style emerged, combining the realism of pre-Islamic India with the formality, elegance, and delicacy of Islamic Iranian art. Flower depictions combined the grace of fifteenth-century Iranian models with India's fascination with nature. At its best, the style strikes a blend of delicacy, exuberance, and restraint.

European art forms began to have an impact during Akbar's reign, adding to the court's already varied aesthetic. Akbar avidly sought European art and allowed his artisans to explore European creative concepts and techniques. For instance, in order to acquire foreign artisan techniques, he despatched a group of craftsmen to the Portuguese town of Goa in 1575. In addition to welcoming European visitors to the court and listening to their opinions, Akbar and his painters also examined the artwork, fabrics, and other items the visitors had brought.

Dutch botanical drawings from the sixteenth and seventeenth centuries had a greater effect on how flowers were shown in Mughal paintings, carpets, and other media than did representations of plants in their native environments. In their paintings, Indian painters began experimenting with European ideas like perspective [11].

A fascinating trend in creating floral pictures from the Dutch Golden Age resulted from geographic discovery and imperial expansion. Numerous coastal regions in South Asia, the Americas, and Africa came under the governmental authority of various Dutch commercial enterprises, which also constructed bases there. The visual language of eighteenth-century European flower illustrations derives from a lengthy artistic heritage of representing natural objects and the most recent advances in botany. It has a realistic but schematic image, which can be seen in the Four Tulips painted by Jacob Marrel, as shown in Fig. 4, which inspired the Patnakalam style of floral painting.

3.1 The Patnakalam

The Mughal heritage provided training for the early Patnakalam artists.

Painters sought support in more recent centres of power when the Mughal ateliers started to decline in the middle of the eighteenth century. British employment

Fig. 4 Four Tulips by Jacob Marrel ca. 1635–45 [12]

and sponsorship were what brought a new hybrid school to Patna artists in the late eighteenth century. The Patnakalam artists came from the Mughal period. When they were employed to do botanical study by their new British masters, they had a deep-rooted artistic heritage. By combining what they had recently learned with what they already knew, a new hybrid creative tradition was born.

A set of botanical images created by Patna artisans for James Kerr is an even older example of drawings created for a British employer. From 1770 to 1782, James Kerr worked as a Scottish Surgeon in Bengal and Bihar as part of the Bengal Establishment. In order to identify and characterize the plants for potential medical or commercial application, Kerr created scientific illustrations of the species. The colour palette in the Kerr images is really vibrant. They exhibit the same poetic delicacy of Mughal blooms at their best. But in many instances, the stiff lines—as though to purposefully warn off pattern-like design—detract from the style, and there's no other visual cue to follow. The meticulousness with which leaf nodes, pistils, and stamens are drawn demonstrates the influence of European botany. According to European Botanical illustration practice, the reproductive portions of the flower are shown individually in some of the drawings [13].

Full-flowered mauve and pale-yellow iris on a short stalk with two sword-shaped, ribbed leaves that curve upward beneath the lowest layer of petals painted by the Patna kalam painters shows a different take on the blue iris by bringing change in the colour palette as represented in Fig. 5 Indian textiles gained increased availability in Europe and the American colonies during the eighteenth and nineteenth centuries. Women from all walks of life wore printed calicos and sprigged muslins, while grand residences in Europe and the colonies were embellished with painted or embroidered bed hangings. Empress Josephine had a picture made in 1809 using a shawl from Kashmir. In less than half a century, Kashmiri shawls and their European

Fig. 5 An Iris flower, eighteenth century [14]

counterparts woven with jacquard became an essential part of women's clothing for both salon and retail settings.

Newer patterns with an emphasis on plants, leaves, stalks, and flowers in their natural bends and curves began to appear during the elegant Shah Jahan period. The features of flowering plants were produced with extreme precision. The study of nature was given additional dimensions by the addition of plants, leaves, sprays, and blooms in harmonious tones, as well as flowery scrolls [15]. The Mughal School of Painting began to develop with the blending of floral studies with a more uniquely Indian flavour. The Wedding coat "Rajputana," as shown in Fig. 6, designed by Richard Cowley has well captured the Mughal florals in which the blue iris is the central theme of the design motif patterns.

3.2 Motifs and Meanings

The diverse array of floral themes that emerged in Mughal art consequently provided fresh designs for the Banaras brocade cloth. Animals of the highest calibre were also created, including tigers, lions, elephants, horses, deer, peacocks, geese, and parrots. The weavers even drew drawings and portraits. This historical group became recognized as Tasveer or pictorial brocades.

All brocade patterns may be grouped under a few headings to make things easier.

Fig. 6 Hand-Painted wedding coat "Rajputana" by Richard Cowley (b.1947) [16]

1. A design pattern with a **Tasveer** or picture motif
2. **Phulwar,** which is the Arabic word for a flowery design
3. **Butidar**, which is Persian for "sprigged floral foliage"
4. A geometric design influenced by Islamic architecture
5. **Shikargah,** which depicts a scene from a hunt

Under several titles, including ***adibel***, ***daurbel***, ***khajuribel***, ***Genda ki bel***, and ***Cane Patti ka bel***, the running patterns of leaves and flowers throughout the foundation were referred to as Bel.

The ***Buti*** (little motif) and ***Buta*** (large motif) motif designs were based on a solitary flower or figure that was not part of a larger pattern. The Latin term Butia, which describes the arrangement of a flower or shrub into a pattern, appears to be the source of the English word Buti. Every Buti variant had a unique name [5].

Ashrafi was a round floral design that mimicked a mohur or gold sovereign. This Buti has a circular design with ***foliage***, ***flowers***, ***buds***, and ***fruits***. The inner circle was frequently filled with a ***chevron*** pattern.

Textile artisans' favourite motif was the green mango, which inspired the design of ***Keri*** Buti. It frequently served as the foundational design and was composed mostly of flowers.

Chand Tara is another popular Buti with the moon and Stars covered all over the ground.

Pan Butā resembled a heart or a betel leaf. This pattern was often found on bridal shawls and scarves [13].

Buti ***jhar-dar*** is a bold butti that consists of a whole branch or sprig. The ***Jhadar*** Buti or a large Buta was the most common motif of Banaras brocades. Jhar consists only of flowers—sometimes realistic and sometimes decorative with a couple of leaves and buds.

A development in complexity was a ***Jungla*** or woodland design consisting of plants, flowers, fruits, birds perching on its branches and butterflies hovering around. It was a complicated and heavy pattern usually used for expensive curtains, quilt covers and floor spreads.

The enigmatic stories of ***Cypress*** Motif symbolize both eternal life and death. We frequently come across references to the Paisley when researching the origins of themes in arts and crafts. However, there is another pattern that stands for both death and endless life; it is more enigmatic and intellectual. And the Cypress is that. Many claim that the cypress tree gave birth to the Paisley when a wayward songbird perched on its top tip, slightly bending it. However, let's start from the beginning. Cypress motifs in Persian carpets indicate that Zoroastrianism persisted in Persia, whether openly or not. The theme is fantastic when utilized in carpets; it frequently covers the full length and is lofty, soaring, and majestic [17].

It typically coexists with the Tree of Life, signifying a similar concept: perpetual life. The pine tree is incorporated into the textiles and fabric through various mediums, as shown in Fig. 7 as kalamkari and in Fig. 8 as the Chintz motif. The Mughals used flower motifs, particularly cypress, in their fabrics. Fascinatingly, French designer Brigitte Singh found old paperwork about these patterns in her father-in-law's archives. Her use of the cypress is the most creative and beautiful application of the theme in contemporary design. The pattern is frequently seen on clothing, household goods, and several items created in India [20].

Fig. 7 Cypress Motif block printed [18]

Fig. 8 Tree of Life motif on chintz [19]

Chikankari needlework with Indian floral designs. White flower embroidery with delicate stitches blossomed during the Mughal court. Many miniatures from the time depict Mughal Emperor Jahangir wearing a flowing white muslin robe embellished with beautiful white embroidery. Empress Nur Jahan, Jahangir's consort with a Persian pedigree and a keen interest in needlework, is supposed to have revived the magnificent craft of flowery white embroidery in Mughal courts. She started a style in the Mughal court by embroidering a hat with Indian blooms using the best white threadwork. Buti (paisley) quickly became the most popular motif in chikankari embroidery, and it is still used today.

The floral patterns on the textiles in the seventeenth century were naturalistic. The use of florals was as a whole unit with stems and leaves. It was used either as a repeating unit or was set singly as a niche for wall hangings and prayer carpets. As the seventeenth century came to an end, the flowers lost their stems and leaves and became more stylized and less realistic. Then they emerged from the realm of inspired creativity. These stylized floral patterns also began to show up in Rajasthani and Gujarati fabrics' rustic embroidery and prints.

The ***Pomegranate***, also known as ***Anaar***, is a motif that has persisted throughout history. From ancient Persia, where they were applied to weapons for defense, to Greece, where the pomegranate calyx crown design became the quintessential shape for royal hair ornaments, history documents their voyage [21]. Like the Persian Paisley and other flower designs, the stylized pomegranate pattern found its way onto textiles and travelled along the historic Silk Road to both the East and the West. Numerous civilizations have used this ancient design theme, which has holy meanings related to fertility, plenty, and our cosmic beginnings.

The ***Lotus*** blossom has been utilized not only for its aesthetic value but also to embody and evoke the eternal attraction of the universe and the simple beauty inherent in objects and mankind.

A prized cultural artifact and ceremonial artifact from the Petaboli region on India's Coromandel coast is **Kalamkari** (south India). With its elaborate blossoming tree and border of garlands and bouquets, this type of Indian mordant-painted cotton fabric (***Palampore***) was specially made for the European market. Their exotic flora was well-liked and a fantastic source of creative stimulation.

The ***Gul-u-butta*** and ***Gul-u-bulbul*** motifs eventually became popular in textiles. Persian poets have used floral and garden imagery, most notably the rose and nightingale, as metaphors for spiritual and earthly love in their epic, lyric, and mystical works, which is a popular theme among Indian couture designers who feature the ***Rose and Nightingale*** in their collections. Gul is a general name for flowers, including rose variations such as ***Gulsurkh*** and ***Gul-i muhammadi***, as well as other sorts of flowers like ***Gul-i sumbul*** (***hyacinth***). The phrase is also used in numerous compound forms, such as ***Gulistan*** (***Rose Garden***), ***Gulnaz*** (flower lover, ***Gulnaz*** (delicate) or ***Gulrukh*** (rose face-young, fresh) [22].

4 Insects, Birds, Animals, and Mythical Creatures

The naturalistic inclinations of the Indian and Mughal weavers and painters are reflected in the abundance of animal themes encountered in miniature paintings and textiles. Frequently portrayed animals include tigers, elephants, deer, birds, and legendary animals like phoenixes. These creatures serve as both ornamental features and symbols. Tigers, for example, represent power and strength and are frequently shown in hunting settings or beside regal people.

Indian art and textiles feature depictions of a variety of animals, including rams, deer, monkeys, squirrels, falcons, peacocks, horses, camels, two-humped camels, hares, cheetahs, and parrots.

Because they could fly and could not communicate, animals and birds were seen as holy and mysterious. They have associations with gods and goddesses, religion, myths, superstitions, and other life- and death-related topics. Man demonstrated his admiration for animals and birds by portraying them in art [23].

Elephants are often used as symbols of monarchy and wealth when seen in processions or conflicts. The elephant is considered an auspicious animal, traditionally connected with monarchy and often represented alongside the goddess of riches, Lakshmi. In temple construction, a row of elephants is frequently seen, as if holding up the entire temple structure as a sign of their protecting strength. The elephant is a popular theme because it represents knowledge, courage, memory, and "Pillaiyar," the elephant god who has temples in every part of South India. A Kanjivaram's pallu, or border, is lined with a row of marching elephants, a gorgeous sight.

A gorgeous graphic panel depicts ***horses*** rushing over the border and a Kanjivaram sari's pallu. This majestic beast is a prominent element of temple mandapams such as yalis, and it also appears in the Vana Singaram, a hunting scene woven on South Indian brocades.

The ***Golden deer***, portrayed in the Ramayana, represents innocence and nonviolence. The peaceful temperament of this animal is brought to life by the silky drape of the Kanjivaram sari. Kanakavalli's characteristic emblem is a graceful, dainty deer that looks stunning on rich silk. It is also associated with Lord Siva's iconography and serves as the chariot for Vayu, the deity of wind.

According to literary and sculptured evidence, the ***Lion*** symbolized Goddess Durga's bright conquering strength (Tejas) at the time of the temple's consecration (kumbabhisheka). A Kanjivaram sari's pallu and border include a lion as a linear pattern.

A ***dragon*** has a long, enormous tail and resembles a creepy beast.

Goblins were a popular cultural heritage that faithfully depicted the Indian terrain in Mughal-era Indian artwork, especially in Northern India and Kashmir. They looked almost the same: a human body, an animal head with horns, and an animal tail.

A purportedly Buddhist stupa in Sirkap, Taxila, is home to the oldest ***Double-headed eagle*** in India. Buddhist caves at Junnar, Maharashtra, have a similar design. The bird is now immortalized in a plethora of temple sculptures and paintings. It first occurs in the South Indian Siva Purana's Satarudra Samhita, where it is depicted as a ***Gandabherunda***, a double-headed legendary bird that swoops down to capture two little elephants in its jaws. The Gandabherunda's stylized feathers fly out from the body and fill the backdrop, while the bottom of the animal, with its wings tucked in a steep dive and its claws gripped, is seen in the foreground. This stunning illustration fills the spaces between the feathers with butterflies, stylized flowers, and birds [24].

In Indian tribal groups, ***Fish*** are powerful emblems of fertility. This design first surfaced in the Indus Valley's archeological records on pottery, Mohenjo-Daro seals, and plaques. Fish served as the dynastic emblem of the Pandyas of Madurai, who lived in the south. The "meenpettu," or borders, of the Kanjivaram sari are decorated with this design. The Vaishnavite-related motifs are also prominent in West Bengali and Orissan textile styles.

Beautiful borders, or bel, are woven in a pattern that matches the pallu or the ground to accentuate the attractiveness of an article of clothing. The boundary is referred to locally as duar or run. One pattern appears repeatedly in a row or constantly in this daur. The elephant border, or hathi daur, and the parrot border, or tota daur, are two designs that have consistently drawn weavers. Some weaves, made entirely of gold without silk, have designs of flowers, mangoes, or creepers. The decorative end is brocaded with green parrots and features gold weft. The lyrical titles of these brocades, such as bulbul chasm (nightingale's eyes), morgala (peacock's neck), and chand tara (moon and stars), are elegant yet fitting [23].

Peacocks (Fig. 9), a prominent theme in Mughal textiles, symbolise the empire's love of extravagance, the natural world, and cultural meaning. Known for their magnificent appearance and brilliant feathers, peacocks were frequently included in the elaborate miniature paintings, fabrics, and ornamental arts of the Mughal era. Given

Fig. 9 Peacock [25]

that peacocks are considered to be symbols of spirituality and majesty in Indian culture, their representation in a variety of artistic mediums emphasized themes of grandeur, immortality, and divine connection. Peacocks were common in hunting trips, gardens, and courtly scenarios; they added to the Mughal artistry's visual majesty and intricate detail [26].

The ***Parrot*** motif is often associated with the narrative of Lord Krishna and Radha, which is a well-recited tale in Hindu sagas and epics. It represents courting and passion. The bird design, which is typically seen on textiles from West Bengal, Rajasthan, and Gujarat, also has romantic connotations and is frequently used in clothing for both brides and grooms.

The ***Goose*** or ***Hansa*** motif pattern, which may be seen on wall paintings in the well-known Ajanta caves in Maharashtra and ceramics from the Indus Valley, is also frequently seen as a symbol of spiritual purity. This design, which has been prevalent in art from before the Mughals, is currently mostly seen on textiles from South India in the form of kasuti stitching and kalamkari prints [27].

5 Shikargah

Shikar *denotes* ***hunting****, and* ***Gah*** *denotes* ***location*** *in* ***Shikargah***

The Shikargah (Fig. 10) scenario first appeared in medieval Persian ornamental arts, miniature paintings, and Mughal architecture before making its way onto textiles. Traditionally influenced by Mughal king hunting, a popular scenario has royals pursuing tigers and deer on horseback alongside elephants(Fig. 11). Slowly, hunting images blended into well-known Banarasi designs, and minor adjustments

Fig. 10 Shikargah on chintz [28]

Fig. 11 Elephant on cotton Sambalpuri saree [29]

to the peak hats worn by British soldiers emerged. The picture was reworked to fit into six yards of cloth, which resulted in the various versions we see today. The Shikargah pattern communicates a tale in addition to being a decorative feature. Every saree transforms into a painting that captures the majesty and drama of royal hunts—a representation of strength, authority, and a close bond with the natural world. Of all the popular designs of Banaras, chintz the Shikargah or arabesque was the finest and has been described as a happy hunting ground Kimkhab which he assumes to have originated in India [30].

Fig. 12 Kimkhab Banaras Brocade [32]

As a symbolic area of the emperor's dominion, the shikārgāh-i muqarrar (favourite and frequently visited imperial hunting plains) is a controlled environment that best captures the emperors' relationship with nature and their perception of their surroundings. It took careful discrimination on the part of the artist to translate these multi-layered ideals of the hunting landscape into visual narratives. He had to keep his identity with the associated textual descriptions while taking into account his narrative mode, his choice of descriptive details, and the prevailing social and artistic conventions.

An allegory that thematically represents the hunt and the hunting ground on an ideological, social, and artistic level is typically the outcome of the composition that shows the emperor engaging in a favourite royal hobby. Crucially, it is a representation of the Mughal worldview as it is filtered via the hunting culture. Keeping these viewpoints in mind, the subsequent section aims to explore the Mughal hunting scene in light of its intricate social, political, and cultural quandaries [31].

Shikargahs can be woven using various techniques, including Kadhwa, Meenadar, Kimkhab(Fig. 12), and Tanchoi. It is estimated that less than 4% of Banaras's overall weaving population now weaves shikargahs. These exquisite weaves require a great deal of support and encouragement for future generations to enjoy them; otherwise, they will all become heritage and museum exhibits.

6 Conclusion

India's cultural and artistic legacy is strongly influenced by nature in its textile themes and pattern creation. Textiles with floral, botanical, animal, and organic features represent important cultural meaning in addition to their visual attractiveness. These natural themes, which range from the lotus and peacock to the paisley and kalasha, have symbolic connotations that link individuals to their daily lives, rituals, and spiritual beliefs. Indian textiles are, thus, a celebration of nature, with symbolism woven into every pattern and design that goes beyond simple decoration. Indian weavers have long been adept at fusing natural elements into their textile creations to produce classic patterns that are both evocative of regional customs and universally appealing. The country's varied landscapes, flora, wildlife, and cultural diversity have all been portrayed via the use of nature-inspired themes in textiles like the exquisite embroidery of Pashmina, the vivid patterns of Kanjeevaram, and the intricate weaving of Benarasi saris. While Indian textile design still draws inspiration from these organic themes today, it also modifies them to fit in with current fashions, fusing age-old artistry with cutting-edge methods. The themes' constant relevance and resonance with audiences both domestically and abroad are guaranteed by this dynamic progression. Indian textiles with motifs inspired by nature are still relevant today, demonstrating the constant relationship between nature and creative expression. Indian textile themes and patterns combine the artistic skill of successive generations with the natural beauty of the land to create a living record of the nation's cultural story. This combination guarantees that Indian textiles will always be a dynamic art form that is profoundly rooted in tradition and contemporary, inspiring and captivating people.

References

1. Maxwell, R. (2012). *Textiles of Southeast Asia: Trade, tradition and transformation*. Tuttle Publishing.
2. Araf. (2024, August 15). *Nature's enduring influence: A look at botanical motifs in modern textile design*. NedGraphics. https://nedgraphics.com/blogs/natures-enduring-influence-a-look-at-botanical-motifs-in-modern-textile-design/
3. Wilson, H. (2011). *Pattern and ornament in the arts of India*. National Geographic Books.
4. Deepshikha, N., Yammiyavar, P., & Nath, N. (2018). Textiles as communicating links for cultural traditions. In *Advances in intelligent systems and computing* (pp. 168–177). Springer. https://doi.org/10.1007/978-981-10-8612-0_19
5. Das, S. (1993). Fabric art: Heritage of India. *Choice Reviews Online, 30*(6), 30–3065. https://doi.org/10.5860/choice.30-3065
6. Victoria and Albert Museum. (n.d.). *The arts of the Mughal Empire V&A*. https://www.vam.ac.uk/articles/the-arts-of-the-mughalempire?srsltid=AfmBOorRH90sY9cB7PZM1JiWGzPB75TU-RkIkqVIuB5QbGmdhmCkM3d-
7. FASHIONING AN EMPIRE: Textiles from Safavid Iran. (2024).
8. Victoria and Albert Museum. (n.d.). *Floorspread | V&A Explore the collections*. Victoria and Albert Museum: Explore the Collections. https://collections.vam.ac.uk/item/O74057/floorspread/
9. Holledge, R. (2018, October 29). *Glorious textiles tell the story of India's past*. Arts Culture—Gulf News. https://gulfnews.com/entertainment/arts-culture/glorious-textiles-tell-the-story-of-indias-past-1.1600519
10. X.com.(n.d.).*X*(FormerlyTwitter).https://x.com/peterfrankopan/status/1304834222822227969
11. Peepultree.world. (n.d.). *Emperor Jahangir: The Naturalist*. https://www.peepultree.world/livehistoryindia/story/living-history/emperor-jahangir-indias-first-naturalist?srsltid=AfmBOop6Bnr1Kj-CeAXU11QS7Etzxt2NNaoM-1U3n04oRBb15rtRCnHz

12. Marrel, J. (n.d.). *Four Tulips: Boter man (Butter Man), Joncker (Nobleman), Grote geplumaceerde (The Great Plumed One), and Voorwint (With the Wind).* The Metropolitan Museum of Art. https://www.metmuseum.org/art/collection/search/337769
13. Schuster, M. (2008). *Field of flowers: Mughal carpets and treasures.* DigitalCommons@ University of Nebraska-Lincoln.
14. *RBSI—An Iris flower 18th century the central painting on the page is of a mauve and pale yellow iris in full bloom on a short stalk with two ribbed sword-shaped leaves curving upwards under the lower.* (n.d.). https://www.rarebooksocietyofindia.org/postDetail.php?id=196174216674_10151140996726675
15. Garg, N. (2021, August 4). *Indian floral patterns in design and textiles.* Marasim—Sustainable Textile, Craftsmanship, Luxury. https://blog.marasim.co/indian-floral-patterns-in-design-and-textiles/
16. Victoria and Albert Museum. (n.d.). *Rajputana | Whittle, Andrew | Cawley, Richard | V&A Explore the collections.* Victoria and Albert Museum: Explore the Collections. https://collections.vam.ac.uk/item/O125173/rajputana-wedding-coat-cawley-richard/
17. Gupta, M. (2021, January 24). *Cypress Motif, the mysterious stories representing death and eternal life.* Marasim—Sustainable Textile, Craftsmanship, Luxury. https://blog.marasim.co/cypress-motif-the-mysterious-stories-representing-death-and-eternal-life/
18. *Wall hanging (Kalamkari) or prayer mat—Works—Museum of Fine Arts, Boston.* (n.d.). https://collections.mfa.org/objects/316769
19. MAP Academy. (2024, March 7). *Tree of life—MAP Academy.* https://mapacademy.io/article/tree-of-life/
20. Dusenberry, M. M., & Bier, C. (2004). *Flowers, dragons, and pine trees: Asian textiles in the Spencer Museum of Art.* Hudson Hills Press. https://ci.nii.ac.jp/ncid/BA74969030
21. Jaipuriar, S. G. (2020). Indian Art and European science: Patnakalam and colonial botanical drawings. *Rupkatha Journal on Interdisciplinary Studies in Humanities, 12*(4). https://doi.org/10.21659/rupkatha.v12n4.14
22. *Selvedge Magazine.* (2022, July 21). Motifs speak: A vocabulary of Indian Textile Motifs. https://www.selvedge.org/blogs/selvedge/motifs-speak-a-vocabulary-of-indian-textile-motifs?srsltid=AfmBOoq5qmnFj64zyuWgI18cGujxZTmrwFRR2AUY4V2f7AD8A-98NCz9
23. Diba, L. (2018). *The rose and the nightingale in Persian art.* Academia. https://www.academia.edu/35622497/The_Rose_and_the_Nightingale_in_Persian_Art
24. *Animal in Indian art—Google Arts & Culture.* (n.d.). Google Arts & Culture. https://artsandculture.google.com/story/animal-in-indian-art-chhatrapati-shivaji-maharaj-vastu-sangrahalaya/OgXRpITH2ZrvJw?hl=en
25. Sandip. (n.d.). *Kalamkari block print seamless pattern, kalamkari, block print seamless pattern.* Dreamstime. https://www.dreamstime.com/illustration/kalamkari.html
26. Kaur, R. (2017). *Peacock motif in phulkari: A comprehensive analysis.* Iitr-in. https://www.academia.edu/30781426/Peacock_Motif_in_Phulkari_A_Comprehensive_Analysis?b=sparse_vector
27. Museum, C. N. S., & Unesco. (2022). *Textiles and clothing along the silk roads: Thematic collection of the cultural exchanges along the silk roads.* UNESCO Publishing.
28. Packer, N. (2012, May 29). *Pattern designs.* Pinterest. https://in.pinterest.com/pin/278801033152322901/
29. *Brown red cotton sambalpuri handloom saree.* (n.d.). GoSwadeshi. https://goswadeshi.in/products/brown-red-cotton-sambalpuri-handloom-saree
30. Kanakavalli, M. (2024, August 27). *Varna Sutra: Mriga, Pakshi—Of Birds & Beasts.* Kanakavalli. https://kanakavalli.com/blogs/kanakavalli-varna-sutra/varna-sutra-mriga-pakshi-of-birds-beasts
31. Parpia, S. (2021). The Artist's gaze: Visual representations of the Mughal hunting landscape. In *Brill eBooks* (pp. 245–283). Brill. https://doi.org/10.1163/9789004499485_011
32. Rana, R. (2024, September 4). *Scenes of a royal hunt: Discover the magnificence of the Banarasi Shikargah.* WeaverStory. https://weaverstory.com/blogs/news/scenes-of-a-royal-hunt-discover-the-magnificence-of-the-banarasi-shikargah

The Business of Textile Design

E. Mokanaasri, Sadhna, and Rajesh Kumar

1 Introduction

Textile design is a very dynamic and multifaceted field. It involves designing patterns and designs for the most varied fabrics. In both apparel and utility sectors, this industry plays a pivotal role, reflecting its broad scope and high demand. It not only includes independent businesses but also integrates the expertise of surface designers, interior designers, embroidery specialists, and textile historians [1]. These experts use their specialized knowledge to innovate and meet market needs, thus reflecting the interdisciplinary nature of textile design. A proper understanding of the field is essential for aspiring designers before venturing into the ever-growing demand for fresh ideas, innovative strategies, and diverse options the market continuously seeks. Modern textile designers must be able to predict market trends and consumer buying behaviours [2], as shown in Fig. 1.

The perception of such foresight emanates from an in-depth understanding of the new paths or directions, sustainable practices, and technological changes that bring many revolutions in various sectors [4]. However, it is pertinent to note that business plans are effective strategies to transform creative ideas into concrete successes. Such a strategy allows designers to associate their products with the ideal customers, to match those creations with market requirements, and eventually achieve commercial success. By understanding these critical points, stakeholders involved in textile design can efficiently work themselves through the intricate nuances of current market scenarios [5].

E. Mokanaasri (✉) · Sadhna · R. Kumar
Department of Fashion Design, School of Arts and Design, Woxsen University, Hyderabad, Telangana, India
e-mail: mokanaadevreddy.98@gmail.com; dr.sadhna345@gmail.com; kumar.rajeshrk2030@gmail.com

Sadhna et al. (eds.), *The Art and Craft of Modern Textile Design*, Sustainable Textiles: Production, Processing, Manufacturing & Chemistry,
https://doi.org/10.1007/978-3-031-86797-2_6

Fig. 1 Business of textile design [3]

This chapter explored these dimensions of textile design. It explains the current market landscape, the changing nature of consumer preferences in textile design, and the strategic approach to business that must be followed to ensure success in this highly competitive domain. It further elaborates on examples of booming textile design industries and their objectives and innovative methodologies. Through such discussion, readers will find valuable insight into the delicate balance between creativity, market awareness, and strategic planning defining textile design excellence.

2 Market Trends in Textile

The global textile market is expected to expand significantly, with projections suggesting a compound annual growth rate (CAGR) of 7.43% from 2024 to 2033. The market is expected to reach an impressive valuation of approximately USD 3767.92 billion by the end of this period. This growth is primarily driven by developments in the Asia-Pacific region, particularly in countries like China, Bangladesh, and India. These countries enjoy bountiful natural resources, rising demand for fashionable clothing and textiles, furniture and furniture accessories, and massive state investments to increase the textile industry [6]. Figure 2 shows the various market trends on a global scale.

The fast development of the fashion industry, along with increasingly popular online shopping platforms, is accelerating the further growth of the market. Companies use social media to connect with existing and potential customers to achieve larger shares in this highly competitive industry. However, the expansion of the global textile market has not been free of time-to-time setbacks. The COVID-19

Fig. 2 Market trends [6, 7]

pandemic was a temporary setback; it hindered supply chains and changed consumer behaviour globally [8].

The adoption of intelligent textiles is another critical growth driver. Advanced fabrics are now made with optical fibres, metals, and conductive polymers, allowing them to interact with their environment. Such innovations are finding applications across industries, including fashion, healthcare, and technology, signalling a transformative shift in textile functionality. Sustainability is also emerging as a central theme in the evolution of the textile industry [7]. This increasing consumer demand for more environmentally friendly products is leading companies to rethink their strategies and prioritize sustainable manufacturing practices. This trend catalyses the shift toward more environmentally friendly production methods, reducing waste and using renewable resources.

Prominent examples of this innovation include DuPont's fur made from plants, changing the face of performance fashion apparel, and Eastman's creative way of reusing discarded carpets into new materials [9]. These innovative ventures show the future of sustainable practices in boosting growth and opening new markets in the global textile industry [10]. With these factors colliding, the future of the textile industry becomes more dynamic, marked by technological innovation, strategic sustainability, and expanding market opportunities. Businesses that embrace these trends and adapt to changing consumer demands will be well-positioned to thrive in the coming decade.

2.1 Technological Advancements and Their Impact on Textile Design

Technological advancement in textile products includes sustainable production processes, environmentally friendly dyes, and intelligent textiles, motivating market expansion due to ecological challenges and customer needs. Technological advancement in textiles is also instrumental in the production of apparel, that is, sportswear, outdoor wear, and functional wear that are marketed to accommodate the wearing comfort of the consumer, sustainability aspect, and relevant fashionable designs [11].

3 Consumer Preferences in Textile Design

Today, consumers are trending to select sustainable and environment-friendly products in the textile design industry. This shift is due to the increasing concern about the environment, where there are positive attitudes towards recycling using green apparel [12]. In this study, it can be observed that consumers are willing to pay a higher price for sustainable clothing and textiles, showing their concern for the environment. It will be helpful to focus on environmentally friendly products that will be appealing to the environmentally friendly generation, including Generation Z [13, 14].

Although sustainability is one of the key influences, product attractiveness and distinction cannot be sidelined within consumers' decision-making process. Consumers seek products that are in harmony with their beliefs and those with unique designs and quality work. When sustainability is incorporated with other factors such as appearance, consumers can be attracted easily by the products manufactured by different brands [15]. Custom creations can also improve consumer's perceptions of value by providing a sense of exclusivity by using sustainable products and that it is fashionable to be environmentally conscious [14].

Technological communication tools such as social networks and digital media have become essential in defining consumer preferences in the textile industry [16]. These platforms enable the quick sharing of trends and brand stories, giving consumers easy access to brands. Among all the given types, it is possible to note that influencer marketing is suitable for promoting sustainable practices and products, as people tend to trust the influencers they follow. Another advantage of presentation on Instagram and Pinterest is the presentation of the unique designs of textiles, which stoke public interest and engagement [17].

However, the choices of textile designs are influenced by cultural factors ranging from fashion to interior designs. When successfully incorporated into textile patterns and designs, these elements can help consumers find a personal touch with the product [18]. Furthermore, home interior design and clothing developments influence the consumer's choice because they want to buy fabrics that suit their lifestyle and part of the world they come from. This relationship between the culture and the design shows just how crucial it is for design to market to meet the demands of the people [19].

4 Business Strategies for Textile Designers

In the textile design industry, adopting advanced technological solutions is essential for enhancing both the creativity of designs and the efficiency of production processes [20]. CAD tools are beneficial for design as they provide high levels of accuracy and creativity for designers, and digital printing technologies can be used to achieve customized and complex patterns in fabrics quickly. Also, new technologies such as automation and data analytics can help minimize costs involved in the production process, eliminate waste and make the production process flexible and competitive, hence gaining more market share [21].

The prominent challenge textile designers face is flexibility in meeting market trends and customer requirements. This entails analysing the market and developing new trends, such as using environmentally friendly and socially responsible materials [22]. These trends show that if designers adjust their products to these expectations, they have a greater chance of providing consumers with designs that will set the brand apart from its rivals. This is not advantageous to the business in terms of increasing customer satisfaction, but it also builds brand equity [23].

4.1 Building Brand Identity and Customer Loyalty

Brand awareness is necessary if textile designers want to build a loyal customer base in the competitive market. It also involves developing a story distinct from the branding of a particular company/organization while attracting its intended market Circle and Branding consistency [24]. Excellent and creative presentation of the products, clubbed with customer satisfaction in terms of guaranteed services, can always build and reinforce the brand's perception in the market, not forgetting complete customer loyalty due to previous efficient experience of a brand's product. Communication can also be enhanced through social media platforms; this means that designers must invest in social media to create awareness and build their brand [25, 26].

Textile designers can adopt several business strategies, including DTC, Wholesale, and Licensing models. The DTC model enables designers to have direct contact with the consumers, thus improving profit margins and allowing close contact with the audience [27]. On the other hand, the wholesale distribution channel involves selling to different distribution centres, such as retailers; this has the advantage of achieving wider market coverage, although it comes with a lower profit margin. License agreements can also allow designers to extend their brand easily without investing in the production process, thus partnering with reputable manufacturers [28].

4.2 *Marketing and Branding Strategies for Textile Designers*

Due to increased competition with other textile designs, marketing and brand awareness are vital to any textile designer. This includes establishing its multi-channel marketing strategy, which includes social media, influencer marketing, and content marketing. In terms of content, marketers should consider designs that capture the attention of their target group and incorporate their designs, space-saving possibilities, and environmentally friendly approaches. Furthermore, strategies such as networking at fairs and fashion events are helpful and may help to get more noticed in the branch [29].

Business management tools such as appropriate pricing strategies and financial forecasting play an essential role in the economic viability of textile design businesses. It is, therefore, crucial for designers to closely examine their cost structure, positioning, and target market when fixing prices. Offering different prices for different customers will also help serve the other classes. At the same time, periodic financial audits will be crucial in determining where to cut costs and where to maximize profits. Knowledge of financial ratios is critical to managing a business as well as to recognizing when it will not be sustainable in the long run [28, 29].

Creativity and innovation are critical success factors that help organizations stay relevant in the textile design industry. Regular research is essential in the qualitative improvement of designs, increased variety of materials and techniques used in the production, and variations in the concept behind designer products. Two external factors are sustainable practices and the use of technology, which also lead to innovation. Through developing a creative and innovative mindset, textile designers can achieve competitive advantages and adapt to the market and the consumers' needs [30].

5 Case Studies of Successful Textile Design Businesses

Case studies of successful textile design businesses help understand strategies, innovations, and best practices that drive excellence. They show how creative vision, smart market positioning, and advanced technologies lead to achievement in sustainable production and intelligent textiles. Adaptability, consumer focus, and strategic partnerships are critical factors for success in this competitive market.

5.1 Case Study 1: Berlin-Based Textile Studio

A textile case study is a Berlin-based project producing engineered knits with a particular interest in Jacquard engineering. In the course of detailed investigations and mapping of colours, a studio delivers fluent gradations and an exciting play on materials and colours. Constantly searching for a technological novelty and a practical result drawn from the digital era connected with traditional production technologies. In addition to its work, the studio also tries to expand the direction of knitting through unique cooperation with other brands and design studios [31].

5.2 Case Study 2: Textile Design Innovation in Hong Kong

A case study showed how design thinking, a human-centred approach, can be applied to develop an innovative concept of home textiles in an elderly centre in Hong Kong. The study followed the five stages of design thinking: the ability to grasp the client's feelings and the definition of the issue, conceptualisation and development of a model, and testing and validation. Here, the goal of the project was to enhance the elderly's living environment in the aspect of textile design that they have never been exposed to before [32].

5.3 Key Insights for Market Success and Sustainable Growth

To succeed in the competitive and dynamic market, businesses need to focus on selecting and promoting high-quality, value-added products that can differentiate them from routine, generic offerings that compete with them. Diversity in operations and flexibility to produce innovative new products further strengthen their market position. Vertical integration is not only a cost-reducing strategy but also lessens the impact of economic fluctuations. This will provide stability as well as resilience. Using design thinking and other human-centred methodologies lets companies find latent opportunities and address unmet consumer needs with creative ideational solutions. This way, established brands and design studios may collaborate to tap into new fields, spread their market, and increase access to an industry landscape that continues to evolve [33].

In today's dynamic market, successful products can only be created by deeply understanding market trends, consumer preferences, and unaddressed needs to ensure that offerings align with the changing demands. Combining traditional craft skills with contemporary knowledge can foster innovation, creating aesthetically appealing and functionally effective designs. Vertical integration and flexible production strategies enhance competitiveness and allow businesses to adapt quickly to changing conditions. Therefore, Utilization of design thinking, which relies on an aspect that develops based on empathy and progressive prototypes and further improves product acceptance by people, increases innovative ideas by providing a fertile ground and unveiling new unexploited potentials, hence ensuring growth in innovation areas for various sectors [33].

6 Conclusion

Challenges include the increased awareness of sustainable textile design using environmentally friendly raw materials and energy-conscious and water-saving technologies. Innovations such as digital textile printing and artificial intelligence adoption are the major trends that positively impact this industry. The market continues to expand with a significant growth rate in the Indian and Chinese markets; there is an increasing social awareness of sustainability, quality, and cultural relevance. Trends and challenges that impact success in the industry have shifted towards the need for sustainability, technology utilization, and both product and service quality and innovation. Brand management and market orientation are two factors that define competitive advantage and sustainable growth in the selected industry. The study concluded that the textile design industry is at a crossroads where embracing sustainability and leveraging technology are not just trends but necessities. Companies that focus on these areas while maintaining high standards of quality and innovation are more likely to secure a lasting competitive advantage and drive long-term growth.

References

1. Studd, R. (2022). The textile design process. *The Design Journal, 5*, 35–49. https://doi.org/10.2752/146069202790718567
2. FibreGuard. (2023b, September 19). *Textile design: The essential introduction for beginners*. https://fibreguard.com/sell/blog/textile-design-the-essential-introduction-for-beginners
3. https://www.freepik.com/free-vector/seamstress-designers-dressmakers-working-tailor-shop_5259942.htm#fromView=search&page=1&position=0&uuid=01d74d68-6b89-4f25-8bc5-6db21e35d118
4. van Erp, T., Carvalho, N. G. P., Gerolamo, M. C., Gonçalves, R., Rytter, N. G. M., & Gladysz, B. (2024). Industry 5.0: A new strategy framework for sustainability management and beyond. *Journal of Cleaner Production, 461*, 142271. https://doi.org/10.1016/j.jclepro.2024.142271
5. Fibre to Fabric. (2007, August). *Indian textile industry—An overview*. https://www.fibre2fashion.com/industry-article/2363/indian-textile-industry-an-overview
6. Precedence Research. (2024, May 22). *Textile market size to worth USD 3,767.92 Billion by 2033*. https://www.precedenceresearch.com/textile-market
7. Júnior, H. L. O., Neves, R. M., Monticeli, F. M., & Dall Agnol, L. (2022). Smart fabric textiles: Recent advances and challenges. *Text, 2*(4), 582–605. https://doi.org/10.3390/textiles2040034
8. Grandview Research. (2024). *Textile market size, share & trends analysis report by raw material (wool, chemical, silk), by product (natural fibers, polyester), by application, by region,*

and segment forecasts, 2024–2030. https://www.grandviewresearch.com/industry-analysis/textile-market
9. Fibre2Fashion. (2017, June 17). *Speakable content*. https://www.fibre2fashion.com/news/company-news/dupont/dupont-named-european-bio-based-materials-company-206408-newsdetails.htm
10. Scherer, G. (2022, April 21). *Sustainable fashion: Biomaterial revolution replacing fur and skins*. Mongabay Environmental News. https://news.mongabay.com/2022/04/sustainable-fashion-biomaterial-revolution-replacing-fur-and-skins/
11. Custom Market Insights. (2024, July 8). *Global textile and apparel market size, share, forecast 2033*. https://www.custommarketinsights.com/report/textile-and-apparel-market/
12. Jimenez-Fernandez, A., Aramendia-Muneta, M. E., & Alzate, M. (2023). Consumers' awareness and attitudes in circular fashion. *Cleaner and Responsible Consumption, 11*, 100144. https://doi.org/10.1016/j.clrc.2023.100144
13. Zeng, L., Moore, M., & Rothenberg, L. (2024). Promoting sustainable consumption: The roles of consumers' domain-specific environmental knowledge and personality traits. *Sustainability, 16*(2), 512. https://doi.org/10.3390/su16020512
14. La Rosa, A., & Jorgensen, J. J. (2021). Influences on consumer engagement with sustainability and the purchase intention of apparel products. *Sustainability, 13*(19), 10655. https://doi.org/10.3390/su131910655
15. Baviskar, D., Chattaraj, A., & Patil, A. (2024). Consumer behaviour and sustainable product choices: Insights from visual trends. *E3S Web of Conferences, 472*, 1–14. https://doi.org/10.1051/e3sconf/202447202002
16. Madhura, K., & Panakaje, N. (2022). Social media as a promotional tool in the fashion industry: A case study on meta platforms Inc. *International Journal of Case Studies in Business, IT, and Education, 6*, 351–365. https://doi.org/10.47992/ijcsbe.2581.6942.0168
17. Goenka, D. (2023, September 16). *How changing consumer trends are shaping the textile industry of the future*. https://www.linkedin.com/pulse/how-changing-consumer-trends-shaping-textile-industry-dipali-goenka/
18. Shin, M. J., Cassidy, T., & Moore, E. M. (2015). Design reinvention for culturally influenced textile products: Focused on traditional Korean Bojagi textiles. *Fashion Practice, 7*(2), 175–198.
19. JD School of Design. (2024, April 17). *Exploring global styles due to cultural influences*. https://www.jdsd.in/exploring-global-styles-due-to-cultural-influences/
20. Han, S., & Yu, S. (2020). Proceedings of the 2020 International Conference on Social Sciences and Big Data Application (ICSSBDA 2020). *Application of Digital Technology in the Design of Textile Products, 484*, 179–183. https://doi.org/10.2991/assehr.k.201030.037
21. The Old Factory. (2024, February 29). *9 Effective strategies for designing your clothing line*. https://theoddfactory.com/9-effective-strategies-for-designing-your-clothing-line
22. Nayak, R. (2019). A review of recent trends in sustainable fashion and textile production. *Current Trends in Fashion Technology & Textile Engineering, 4*(5), 555648. https://doi.org/10.19080/ctftte.2019.04.555648
23. Data Bridge Market Research. (2023, August 1). *Textile market size, value, statistics, & industry analysis by 2030*. https://www.databridgemarketresearch.com/reports/global-textile-market
24. Lane Keller, K. (2009). Building strong brands in a modern marketing communications environment. *Journal of Marketing Communications, 15*, 139–155. https://doi.org/10.1080/13527260902757530
25. Paromita. (2024, March 31). *Brand identity design process for apparel startups*. Fashion Tech Pack Designer. https://www.paromitadas.com/clothing-designer/brand-identity/
26. *How can you use textiles to create a consistent brand identity for your fashion designs?* (2023, September21).https://www.linkedin.com/advice/0/how-can-you-use-textiles-create-consistent
27. McKee, S., Sands, S., Pallant, J. I., & Cohen, J. (2023). The evolving direct-to-consumer retail model: A review and research agenda. *International Journal of Consumer Studies, 47*(6), 2816–2842. https://doi.org/10.1111/ijcs.12972

28. Digital Transformation Skills. (2023, April 18). *10 inspiring textiles business model transformation examples*. https://digitaltransformationskills.com/textiles-business-model-transformation/
29. Valades, B. (2023, June 14). *The best marketing strategies for fashion designers | BluCactus*. BluCactus Digital Marketing. https://blucactus.blue/marketing-strategies-for-fashion-designers/
30. Ali, I. (2023, December 7). *Effective strategies for success in textile manufacturing*. https://www.linkedin.com/pulse/effective-strategies-success-textile-manufacturing-ali-chinioti-mjxwf/
31. Case Studies. (2024, March 4). *Case Studies—Berlin based textile studio focused on engineered knits*. https://thecasestudies.com/
32. Yao, M., & Li, L. (2022). Design thinking applied to home textiles innovation: A case study in an elderly Centre in Hong Kong. *Designs, 6*(3), 49. https://doi.org/10.3390/designs6030049
33. Malik, V. (2023, June 1). *How to do business analysis of textile companies*. https://www.drvijaymalik.com/textile-companies/

The Role of Textile Design in Fashion

P. B. Shree Sahana

1 Introduction

Textile design is an essential and rapidly growing segment within the fashion business and plays a crucial role in determining the fashion and functional characteristics of clothing. It is an art form and technical profession concerned with fabric construction and innovations in shaping and contributing to apparel's tone and tactile and aesthetic appeal. This area of design is crucial and is at the heart of dictating what the consumer should prefer, setting trends, and defining the identity of fashion brands [1].

Indeed, it transcends ornamentation and becomes a link between the artwork's aesthetics, the material's mechanics, and clothing functionality [2]. At the intersection of creative design and technical expertise, textile design includes multiple stages, from the idea of patterns and texture surface up to the degree of fibres chosen to the Materials Response characteristic in the manufacturing process [3].

Fashion design is most intimately connected to textile design because it involves selecting fabrics that will be used to bring a designer's concept into realization. Closely related to fashion designers, textile designers can collaborate on fabric selections that correspond to the silhouette and purpose of their collections. Indeed, fashion is dynamic, and textile designers always meet new trends and add new qualities to fabric surfaces and forms through aesthetic concerns and technological developments [4].

P. B. Shree Sahana (✉)
Department of Fashion Technology, National Institute of Fashion Technology, Chennai, Tamil Nadu, India

Sadhna et al. (eds.), *The Art and Craft of Modern Textile Design*, Sustainable Textiles: Production, Processing, Manufacturing & Chemistry,
https://doi.org/10.1007/978-3-031-86797-2_7

2 Historical Context

Generally, changes in society, cultures, and technologies bring changes in textile design. Crowdsourcing forms textiles as objects and the aesthetic and cultural aspects of minstrelsy are illuminating, as is the growth of ornament as an art form. For instance, linen was sought in ancient Egyptian civilization because of its lightweight, elaborate likeness and connotations, which affirmed status and religion [5].

During the medieval period, texts like tapestries, for instance, depicted history, religion, and royalty, and the engravings on the walls served as records of the societies' cultures [6].

Socio-cultural change regarding textile production was made during the Industrial Revolution in the eighteenth and nineteenth centuries. Inventions like the Jacquard loom and power loom led to increased production speeds. This was the start of the move from handcrafted finished products to more machine-produced items in textiles that people of all classes could afford [7].

The innovations of the past century in synthetic fibres like nylon and polyester have made them cheaper and better substitutes for natural fibres in fabric manufacturing. These synthetic fibres improved new textile processes, such as digital printing, and combined them to produce fabrics that offered better functionality and sturdiness than the original forms [8].

Due to the development of many technicalities in the middle of the twentieth century, fabric types that could be used for performances could also be produced for particular usages, like workout dresses and heat-bearing materials for industrial purposes. These innovations integrated fashion with science and technology, giving birth to functional textiles in the fashion industry.

Most recent advances in smart textiles have shifted to the next level. The fabrics now contain technologies that allow them to perform a particular task while working under or responding to certain conditions. Hitherto, smart textiles came embedded with health-monitoring sensors, fibres that adapt to the body temperature, and fabrics that wick moisture for a better feel. Some samples include bright clothing that can communicate information and innovative clothing that can generate power. These colourful clothes can clean themselves or have anti-bacterial qualities. Such enhancements open horizons for fashion appearance, sportswear, the medical field, and even a military uniform, illustrating the transformative potential for technology incorporated in textile design [9].

3 The Role of Textile Designers

Textile designers are critical in determining the fabric to use since aesthetic value and functionality are essential. Their roles are complex and encompass design, identification of suitable materials, and other aspects concerning each textile in terms of application, appearance, performance, and market. This includes understanding various elements such as feel, wear resistance, or elasticity depending on the type, character of the fabric, etc., which are prerequisites of making cloth fit for wearables, home furnishing, and other applications in fashion and even technology like car interiors or medical garments. Knowing some basics of colour theory is vital because it helps designers develop lovely patterns and colour combinations that consumers will like and reflect the trends of the current season. Besides, textile designers should also consider the market trends, as well as the needs of consumers, to make sure that the products meet aesthetic and functional aspects [10].

3.1 Responsibilities

Textile designers work under the pressure of designing fabrics for both looks and performance; as such, they need to have a good grasp of fabrics when designing and constructing them. This requires them to consider the fibre's texture, weight, durability, and stretchability. The patterns that should be adopted in the development of the textile should also consider that the textile is a practical product and an aesthetic one [11]. For instance, fabrics utilized for formal wear require qualities such as aesthetic appeal. In contrast, unlike those intended for athletic wear, they need features such as ventilation and no dampness retention [12].

Designers must also continually be aware of the market, consumer feedback, and even the applicable system of colour. They must also be able to predict market trends and use this understanding to shape their designs to fit the market [11]. The role of a textile designer is to achieve creative and aesthetic harmony while paying attention to the functionality of each fabric and the cohesiveness of the fashion collection. Thus, the role of a textile designer is all about order and creating balanced fashion collections while preserving the creativity of a fashion fabric.

3.2 Collaboration

Another key role of a textile designer is to participate in cross-functional teamwork since most work hand in hand with fashion designers, speculating the outfit to associate with their fabric designs. This cooperation commences at the concept formulation stage, where the textile designer should ensure their designs are in harmony with the fashion designer's idea, temperance, and tone. There is mutual interdependence between the two because garment manufacturers should clearly explain the fabric characteristics and their specifications to the textile designers. Once they have developed the fabric designs, they consult and negotiate on the selection of the materials within the contexts of the duration and cost within which the textiles should be produced. Collaboration can also be observed between manufacturers and suppliers, with which the textile designers need to consult to make the fabric effective and high-quality. This means knowing the dimensional advances of weaving, dyeing, and finishing techniques, to name a few. Still, a few are based on the competencies in textile designs and responding to these competencies by making specific changes as needed based on the production partners with whom they deal [10].

3.3 Innovation

Creativity is the driving force of fashion textile design. Designers continuously search for new fabrics that satisfy changing visual and practical needs. This desire to be unique motivates textile designers to increase their creativity and create more inventive and superb fabrics through weaving and dyeing techniques [13].

Lately, technology has been instrumental in the process due to the emergence of smart textiles. These textiles weave technology into the fabric of the textiles, for instance, sensors for capturing health information or conductive yarns for texturizing apparel and clothing that could interface with electronic equipment. Furthermore, sustainability as a concept emerged as relevant to fashion, as the worn items are extended conceptions of material and production liability that have led to research and discovery of sustainable material, recyclable fibres, experimenting biodegradable or recycled products, and reduction in wastes in the process of production. Thus, textile designers have been designing fabrics that cut across fancy technologies and sustainability, promoting and revolutionizing modern fashion trends.

4 Techniques and Processes in Textile Design

Textile design encompasses various techniques and processes for innovating fabrics, aesthetics, and utility. The diversity of the methods applied in textile design is vital in creating an immense amount of textile materials utilized in fashioning body wear, housing, and other uses. Knowledge of these techniques helps designers control the material properties to produce fabrics that meet the targeted design objectives, the intended consumers, and the evolving technologies [14].

4.1 Types of Textile Design

The craft of textile design is the intricate weaving, printing, dyeing, and embroidering of textures and patterns onto fabrics. In fashion and interior design, textile design is instrumental in shaping aesthetic and functional properties, driving trends, creating visual appeal, and enhancing the tactility of clothing, upholstery, and home decor elements.

4.1.1 Woven Design

Woven design is one of the most basic techniques in designing textiles: two types of yarns—warp and weft—are interlinked at right angles to form a fabric. It involves the manipulation of yarns to give different weave structures: plain, twill, satin, etc. Each pattern yields different textures, drapes, and wearability. Due to their material density, woven patterns are applied to fabrics such as denim, tweed, and silk to fit specific uses. Heavier yarns with tighter weaves give better strength, and a stiffer cloth, often used in suits or upholstery, and more flowing yarns woven loosely are best for summer wear or scarves [15, 16].

4.1.2 Printed Design

The second important method is printed design, which means applying patterns to the fabrics and applying techniques such as screen printing, digital prints, and rotary prints. While screen printing is most effective for simple designs that use two or more colours, digital printing picks up detail and multiple tones in the graphics for less money. Printed textiles include simple designs from the most intricate and photographic material, hence applied in fashion and home textiles. The most important factors considered when choosing the desired printing method are the production size, fabric type, and design [17].

4.1.3 Mixed-Media Design

Semi-abstract/mixed-media design is an innovative technique, whereby the designer can combine any material and method to produce unique textures. This may involve combining woven and printed parts or adding non-fabric such as beadwork, sequins, or metal thread. Mixed media designs have become a culturally standard form of design within the new generation of fashion, where designers combine conventional upholstery fabrics with other materials to achieve a culturally unique touch. The above approach offers many creative options that can create fabrics that are as unique and kinetic as one could imagine [17]. Figure 1 shows examples of different types of textile design fabrics.

4.2 Design Tools

CAD is the most accurate and efficient tool, and it has now been included in textile design. Designers can quickly draw and create fabric designs and experiment with fabrics. CAD features allow designers to create representations of patterns and textures, changing one repeat to colours and other fabric characteristics without altering the original fabric [19].

Fig. 1 Examples of woven design fabric, printed design fabric, mixed-media design fabric from left to right [18]

This allows an instant tactile feel of how the design will look on different fabrics and increases precision. One benefit of CAD over other modelling design tools is its capability to represent any fabric type's fuss, texture, and elasticity without needing a physical prototype. CAD also has benefits in revising designs and considering design changes and is therefore particularly useful for a quickly changing market such as fashion. Moreover, equipped with CAD, the designer can easily link it with other computer-assisted technologies, including pattern-making software and textile production equipment, so that the final product is the correct design of the heart. Such integration becomes essential in achieving near-photorealistic design accuracy as the trend towards personalized designs and shorter production runs escalates [20].

4.3 Material Knowledge

For textile designers, it is vital to understand the diverse fabrics that are therefore available as well as their characteristics since the choice of fabric would dictate form as well as use. Natural fibres like cotton, silk, and wool, as well as synthetic fibres like polyester, nylon, acrylic, etc., all have characteristics that influence their ability to absorb dyes and prints, production methods, etc. For instance, cotton is soft, lightweight, and calm, suitable for casual clothes, while shine is associated with silks that make them ideal for formal wear. Wool is warm, so it is perfect for outer garments. Synthetic fibre, such as polyester, is also strong, easy to wash, and very good for sportswear [21].

It is equally essential for designers to be aware of fibre characteristics used during dyeing, weaving, and finishing; some are known to shrink or elongate upon exposure to heat or humidity. Besides, with the aspects of sustainability becoming relevant, the focus on such materials as organic cotton, Hemp, and recycled polyester when creating the outfit so that owners and designers would consider fashion that is okay with ethical and environmental standards in fashion [22].

5 Trends and Market Research

Trend forecasting is essential in textile design because it helps designers respond proactively to consumers' expectations. Through research, designers can estimate future trends in the textile industry through trends such as cultural, social media, and past trends. It also guarantees that designers develop textiles that reflect the consumer's taste and preference, which may be changed by society, seasons, or other factors [23].

To research the market needs, designers use several approaches, such as surveys and focus groups, which help ascertain the market's attitude, desire, and expectation towards certain products. These tools aid designers in measuring the perception of particular styles, colours, or materiality and point to potential developments. Closely reading industry reports and analysing sales data also help to determine which trends are rising and which are declining, which is useful when further developing the concept. Further, social media is a tool that can be used to determine current tendencies among consumers; it helps designers to respond to new challenges. In this way, the specified research techniques help develop relevant, market-oriented products that consumers will appreciate and allow textile designers to remain relevant in the competitive industry.

6 Sustainability in Textile Design

Sustainability issues have emerged as a critical aspect of textiles as the fashion and textile industries continue to expand and look for ways to reduce their effect on the environment. Eco-friendly practices in textiles include reducing wastage, using energy, and using environmentally friendly products in production [24].

6.1 Use of Eco-Friendly Materials

One of the essential practices in sustainable textile design is promoting environmentally friendly fabrics within the textile manufacturing process, including organic cotton, Hemp, and recycled fibre, among others. These materials usually require fewer pesticides, less water, and less energy than conventional fabrics. Also, recycled fibres are employed; for example, polyester is derived from used bottles, which prolongs the time such materials spend in landfills [25].

6.2 Efficient Production Methods

Sustainable manufacturing aims to trim waste and energy use in product design and development stages. One such approach is zero-waste pattern making, where the intention is to minimize the fabric waste to none. The second innovative approach in sustainable production is waterless dyeing technologies, where fabrics are dyed in supercritical carbon dioxide. This also reduces water usage and pollution by providing a better solution than the normal dyeing process, which requires a lot of water and chemicals. In addition, manufacturers are also using less energy and emitting less CO_2 by installing better energy-efficient devices and processes [26].

6.3 Circular Fashion and Recycling

Circular fashion is another innovative model now under development to return fashion items easily disassembled, recycled or repurposed when entirely worn. Those discerning processes also effectively use recycling and upcycle strategies in the design cycle so that the demand for new products does not increase and the textile products, which could otherwise end up in landfills, do not. Upcycling can be defined as taking used garments or materials and giving them new forms and uses. In contrast, a closed-loop recycling system now attempts to keep the fabrics in use for as long as possible by recycling, remanufacturing, Reducing Waste and using resources optimally. This means making a cycle that can be continuously run, in which the textile can be used again and again without total depletion [27].

7 Career Opportunities in Textile Design

Opportunities for textile designers are promising. They can work as fabric designers, design assistants, or product developers in and for larger fashion companies. Qualifications necessary to fulfil the positions are creativity, computer and design skills, knowledge of fabrics, and knowledge of the market.

A fabric designer works on the patterns, touch, and feel of fabrics bought for fashion, home decoration, and other industries. Design assistants are in charge of supporting senior designers with their research, the building of prototypes, and finally translating concepts into samples in fabric. Product developer works with apparel designers, makers, and suppliers to achieve the design of fabrics that are both fashionable/functional and, most importantly, 'doable' from a production perspective. Retail technologists specialize in evaluating fabrics with respect to their strength, quality, and other characteristics required by the industry. As sustainability awareness extends across the textile market, sustainable textile designers design garments using natural resources and processes of sustainable development goals.

CAD technicians and digital textile designers employ computer software wares, including computer-aided design CAD, to design fabric prints, which can be produced digitally or through automated systems.

In the same way, textile merchandisers study the market and customer behaviour in order to design textiles that meet current tastes and needs. Textile colourists also ensure that the right colour is used appropriately on the textile material for functions such as manufacturing processes and looks. A textile production manager supervises the production process of textile products, including style, quality, and specified measurements. Last but not least, textile researchers are involved in studying new materials, new fibres, advanced technologies in textiles and the production of new textile products. These positions demand creativity, technical skills and market insight and constitute the central spine of the textile design market niche [28, 29].

8 Conclusion

Textile design is essential in defining the fashion industry's yesterday, today, and tomorrow. This was achieved by identifying the factors that have shaped the evolution of the product; the culture, technology, and designers discuss the dynamism of the product. Textile design has a social aspect within the market, which, over the years, was characterized by hand-woven fabrics to industrial technologies that make fashion accessible to anyone. Today, it maintains that tradition and brings together traditions of craftsmanship and innovations to the needs of the twenty-first-century market. However, sustainability, ethical production, or constantly changing consumer demands are burning issues in the textile industry, which give even greater significance to textile design. Finally, textile design is the most promising and, at the same time, responsibly positioned to address historic challenges as well as explore innovative opportunities. Fashion cannot be reduced to nor defined by craft or discipline; rather, it is the lifeblood of fashion and a formative historical concept through which the acts of envisioning, making, and wearing clothing are understood. In addition to playing an aesthetic role, the evaluation of textile design presents the values of society, the development of technology, and the constantly evolving desires of the people. This insightful dualism requires creative and systematic articulation of how textile design remains thought-provoking, innovative, and on the vanguard of spearheading the fashion industry to a greener and brighter future.

References

1. FibreGuard. (2023, September 19). *Textile design: The essential introduction for beginners*. https://fibreguard.com/sell/blog/textile-design-the-essential-introduction-for-beginners
2. Malik, B. S., & Azhar, N. (2015). Role of inspiration in creating textile design. *International Journal of Engineering Research and Applications, 5*(5), 1–7.

3. Studd, R. (2002). The textile design process. *The Design Journal, 5*(1), 35–49. https://doi.org/10.2752/146069202790718567
4. Lisaadelhi. (2023, May 26). *Difference between fashion & textile design*. Leher School of Design. https://www.lsod.in/difference-between-fashion-textile-design/
5. Un-Hjdm. (2024, September 30). *The evolution of textile design: From ancient techniques to modern innovations*. JD Institute of Fashion Technology. https://www.jdinstitute.edu.in/the-evolution-of-textile-design-from-ancient-techniques-to-modern-innovations/
6. Campbell, T. P. (1 C.E., January 1). *European tapestry production and patronage, 1400–1600*. The Met's Heilbrunn Timeline of Art History. https://www.metmuseum.org/toah/hd/taps/hd_taps.html
7. Farago, A. (2017, October 18). *The textile industry during the industrial revolution*. globalEDGE. https://globaledge.msu.edu/blog/post/54483/the-textile-industry-during-the-industrial-revolution
8. Content Team. (2024, November 11). *Natural vs. synthetic fibers: What's the difference?* Urban Space Home Furnishing. https://www.urbanspacestore.in/blogs/news/natural-vs-synthetic-fibers-key-differences
9. Dufossé, L. (2024). New research trends for textiles, a bright future. *Text, 4*(3), 356–358. https://doi.org/10.3390/textiles4030021
10. Kumar, S. (2018, December 20). *Role of textile designer—Style2Designer.com*. Style2Designer. https://style2designer.com/textile-techniques/role-textile-designer/
11. Paromita. (2024, September 4). *Textile designer: Unveiling Artistry from Inspiration to Innovation*. Fashion Tech Pack Designer. https://www.paromitadas.com/clothing-designer/textile-designer/
12. Nasrin, S., Mandal, S., Islam, M. M., Petrova, A., Agnew, R. J., & Boorady, L. M. (2023). Factors affecting the sweat-drying performance of active sportswear—A review. *Text, 3*(3), 319–338. https://doi.org/10.3390/textiles3030022
13. Valentine, L., Ballie, J., Bletcher, J., Robertson, S., & Stevenson, F. (2017). Design thinking for textiles: Let's make it meaningful. *The Design Journal, 20*(Suppl. 1), S964–S976. https://doi.org/10.1080/14606925.2017.1353041
14. Hossain, M. T., Shahid, M. A., Limon, M. G. M., Hossain, I., & Mahmud, N. (2024). Techniques, applications, and challenges in textiles for a sustainable future. *Journal of Open Innovation Technology Market and Complexity, 10*(1), 100230. https://doi.org/10.1016/j.joitmc.2024.100230
15. Shenton, J., & Ridsdale, E. (2014). *Woven textile design*. Laurence King Publishing. http://public.eblib.com/choice/publicfullrecord.aspx?p=1876179
16. Victoria and Albert Museum, & Rothstein, N. (1994). *The Victoria & Albert Museum's textile collection: Woven textile design in Britain from 1750 to 1850*. Canopy Books. https://search.worldcat.org/title/31203715
17. Clarke, S. (2011). *Textile design*. Laurence King. http://www.credoreference.com/book/lkingtd
18. www.freepik.com/
19. Bhanushali, M. (2023, June 6). *5 Ways computer-aided design benefits the textile industry*. Textile Value Chain: Textile Magazine, Textile News, Apparel News, Fashion News. https://textilevaluechain.in/in-depth-analysis/articles/5-ways-computer-aided-design-benefits-the-textile-industry
20. Indrie, L., Diaz-García, P., & Kazlacheva, Z. (2019). The use of CAD/CAM for textile designs and fabrics. *Applied Researches in Technics Technologies and Education, 7*(1), 24–28. https://doi.org/10.15547/artte.2019.01.003
21. Textile-Engineering. (2023, May 5). *Natural and synthetic textiles: Advantages and disadvantages*. https://textileengineering.net/natural-and-synthetic-textiles/
22. Marr, A., & Hoyes, R. (2016). Making material knowledge: Process-led textile research as an active source for design innovation. *Journal of Textile Design Research and Practice, 4*(1), 5–32. https://doi.org/10.1080/20511787.2016.1255447

23. Sodhi, A. (2016, May 3). *The significance of fashion and trend forecasting in design development.* Fibre2Fashion. https://www.fibre2fashion.com/industry-article/7734/the-significance-of-fashion-and-trend-forecasting-in-design-development
24. Muthu, S. S., & Gardetti, M. A. (2020). Sustainability in the textile and apparel industries. In *Sustainable textiles*. Springer. https://doi.org/10.1007/978-3-030-37929-2
25. Kumar, P. S., Yaashikaa, P. R., & Carolin, C. F. (2021). Sustainability in textile design. In *Sustainable textiles* (pp. 39–51). Springer. https://doi.org/10.1007/978-981-16-2466-7_3
26. Zhang, Y., Kang, H., Hou, H., Shao, S., Sun, X., Qin, C., & Zhang, S. (2018). Improved design for textile production process based on life cycle assessment. *Clean Technologies and Environmental Policy, 20*(6), 1355–1365. https://doi.org/10.1007/s10098-018-1572-9
27. NetZeroCities. (n.d.). *Textile and fashion design for circularity.* https://netzerocities.app/resource-2301
28. Fibre2Fashion. (2013, May 30). *Occupational undertakings & careers in textile designing.* https://www.fibre2fashion.com/industry-article/6920/submitArticleLink
29. JD Institute of Fashion Technology. (2024, October 11). *Career opportunities in textile design.* https://www.jdinstitute.edu.in/career-opportunities-in-textile-design/

Challenges and Opportunities in Modern Textile Design

S. Aishwariya and S. Greeshma

1 Introduction

Textile design studies the fabric's structures, patterns and textures. The field of textile design is integral to aesthetics and functionality that impact various industries, from haute couture to interior decoration. 3D weaving, tapestry weaving, colour changing patterns, olar curtains, smart door mats, and grass have been made into tote bags adorned with embroidery, textile-based wallpapers, tile/floor stickers, 3D printing; and many more technologies are integrated with the textile industry. Textile design is more from textile bases to many other surfaces and mediums that need a motif/pattern to enhance their value and create an interest for the eyes. Colour and design play a key role in designing and are the core elements of learning in textile design [1].

The methodologies and materials of textile design have changed, yet its core is still embodied in society's culture, artistic, and practical requirements. The textile design industry has undergone immense transformations in years that the dominant fashion trends and technological developments have primarily driven. Textiles were labour-intensive, associated with skilled craftsmanship and ancient techniques such as weaving, dyeing, and embroidery. The basic building blocks were natural fibres such as cotton, wool, and silk, emphasizing beauty and durability [2].

Today, various synthetic materials are employed to meet the diverse demands of modern life, including applications in protection, mobility, construction, healthcare,

S. Aishwariya
Department of Textile Design, National Institute of Fashion Technology,
Chennai, Tamil Nadu, India
e-mail: aishwariya.s@nift.ac.in

S. Greeshma (✉)
Department of Fashion Design, School of Arts and Design, Woxsen University,
Hyderabad, Telangana, India

Sadhna et al. (eds.), *The Art and Craft of Modern Textile Design*, Sustainable
Textiles: Production, Processing, Manufacturing & Chemistry,
https://doi.org/10.1007/978-3-031-86797-2_8

agriculture, aerospace, underground mining, and fire-resistant clothing. These textiles are crucial for adapting to society's fast-paced and evolving needs. For example, Olympic swimmers utilize a biomimetic fabric that mimics fish skin to enhance their performance in the water. With the rise of social media and influencer culture, travel has gained immense popularity; thus, the textiles designed for adventure sports and travel gear are crafted to balance aesthetics with functionality and performance [3].

2 Current Trends in Textile Design

In the realm of textile design, three-dimensional printing has emerged as a groundbreaking innovation. It allows designers to produce elaborate and detailed patterns and structures that conventional production techniques struggle to replicate [4]. A notable illustration of this phenomenon is observed in the oeuvre of fashion designer Iris van Herpen, who has seamlessly incorporated 3D-printed components into her haute couture collections, yielding garments characterized by elaborate and sculptural designs that subvert traditional fashion conventions [5].

In a parallel vein, Nike has harnessed the capabilities of 3D printing in the formulation of its Flyprint footwear, employing the technology to produce lightweight and breathable textiles tailored to individual athletes' specific needs, thereby augmenting both performance and comfort [6]. Beyond this, 3D printing is applied to achieve remarkable functional and artistic textiles. Stratasys, in partnership with designer Ganit Goldstein, has innovated 3D-printed wearable art pieces, amalgamating flexible materials with sophisticated geometries to expand the frontiers of textile design [7].

Furthermore, Danit Peleg's entirely 3D-printed fashion collection underscores the potential of this technology for sustainable and decentralized production, enabling designers to fabricate garments from their residences while minimizing waste and resource consumption [8].

Computer-aided design (CAD) applications facilitate crafting complex designs with remarkable exactness. This has been proven to create designs, patterns, motifs, and theme boards in a few seconds, which helps increase production with more precision. The advent of CAD/CAM solutions, characterized by contemporary designs and innovative colour palettes, significantly augment textile fabrics' aesthetic appeal and competitive edge, thereby addressing the swiftly evolving fashion requirements on both national and international scales [9].

Within the handloom sector, this development facilitates value enhancement, thereby increasing income and fostering social empowerment for artisans and enabling them to compete effectively within global marketplaces. Today, in our country, it is ironic that 90% of all weaving clusters employ CAD to design patterns and punch cards for weaving. The entire process is digitalized. The weavers also operate the power loom using CAD/CAM solutions, which are noted for their operational efficiency, user-friendliness, and economic viability; it is revolutionizing both handloom and power loom industries, with particular emphasis on textile design methodologies.

Smart textiles, another advancement, which integrate electronic components into fabrics are gaining traction, facilitating garments that can monitor health or undergo colour changes. Indeed, athletes training for competitions larger than life like the Olympics need more gadgets to support and monitor their training and performance. In this context, body monitoring garments have a considerable role, and their application has also extended to hospital textiles. Body monitoring garments represent advanced wearable technologies integrated with sensors to track physiological metrics such as heart rate, body temperature, and movement dynamics [10, 11]. The utilization of these garments is progressively becoming prevalent within academic research, particularly in domains such as sports science, medicine, and ergonomics. Within the sports field, they provide immediate feedback systems that aid in optimizing athletic performance while lessening the potential for injuries.

In the healthcare, these garments support ongoing patient monitoring, which is instrumental in managing chronic diseases and investigating rehabilitation strategies. Within occupational studies, they serve as tools for evaluating physical strain and refining workplace ergonomics. These garments provide significant, non-invasive data collection capabilities, propelling research initiatives across many academic disciplines [12].

Coherently, environmentally sustainable methods are surfacing as an important trend in design, with creators prioritizing eco-conscious materials, products and technologies in the forefront. New fibres like hemp, linen, sisal, agave, ramie, raffia, pine, aloe, milk, corn, orange peel, citrus leather, organic cotton, recycled nylon, and polyester are used. The water pollution caused by textile processing is given keen importance, and unbleached, undyed, naturally dyed, eco printing, painting with natural dyes is now used in sleepwear, kids wear, bridal, and couture.

Regarding prospects, the trajectory of textile design is poised to be characterized by the confluence of Fashion and technology. Developments, including 3D manufacturing, nano-engineering, and AI, will likely significantly impact the commercial arena. Fabrics produced through 3D printing could yield garments customized to individual body dimensions, thereby minimizing waste and enhancing personalization. The evolution of smart textiles is expected to continue, integrating more sophisticated sensors for interactive apparel. The principle of sustainability will persist as a central tenet, with advancements in the creation of biodegradable fabrics and closed-loop recycling systems. The industry shall persist in harmonizing aesthetics, functionality, and ethical considerations, thereby expanding the potential of textiles.

3 History of Textile Design

Textile design possesses a rich historical lineage spanning thousands of years. In order to fabricate functional textiles, early humans utilized animal pelts, natural substances such as cotton and flax, and rudimentary techniques such as felting and weaving. After hunting, the bones of the animals were sculpted to make a hole resembling the modern-day sewing needle, and hides torn into thin strips were used to attach the pieces. The application of colour and sketching of visuals around them as painted illustrations was common [13, 14].

Ancient Egyptians employed advanced weaving methodologies to produce exquisite linen [15]. The civilization of the Indus Valley is renowned for its use of cotton [16], while China innovated weaving and silk production techniques that significantly altered the landscape of international trade. The Medieval period was characterized by a pronounced specialization in the domain of textile production, with various regions developing their own unique stylistic expressions. The Renaissance heralded an era of heightened artistic engagement in textile design, during which luxurious fabrics, intricate tapestries, and elaborate embroidery were explicitly crafted for the elite social classes. The renown of Italian, French, and Spanish textile workshops increased. The era of the Industrial Revolution, spanning the eighteenth and nineteenth centuries, caused a profound alteration in how textiles were produced. The advent of mechanized looms and the establishment of factories significantly accelerated the speed and volume of production. Furthermore, innovations in dyeing and printing techniques facilitated a remarkable expansion of printed textiles during this period.

4 Cultural Representation Through Textile Design

Textile design strongly bonds culture, often revealing a society's identity, traditions, and values. Since historical times, textiles have represented power, status, and spirituality. One fine example is silk, which has represented high status and prestige for more than 5000 years in China. Skilled artisans who perfected techniques in silk production also presented elaborate designs in both the Song and Ming Dynasties, such as designs of dragons and phoenixes. The Silk Road was essential in the history of global trade. It distributed the influence of silk worldwide. Now, the qipao or cheongsam is a symbol of elegance in silk.

The Japanese textile traditions are a treasure trove of patterns that reflect nature and seasonal themes, including kimono fabrics and techniques like shibori or tie-dye. Shibori, kasuri, and yuzen (Fig. 1) [17] emphasize Japan's deep connection with nature through delicate designs [18]. In contrast, Islamic textiles, such as Ottoman rugs (Fig. 2) and Persian silk brocades (Fig. 3), adhere to religious guidelines by focusing on geometric patterns and calligraphy. These textiles are

Fig. 1 Kosode with Yuzen [17]

Fig. 2 Ottoman period—Classical Turkish Carpet [19]

Fig. 3 Persian Silk Brocade [20]

Fig. 4 Tupa-inca-tunic [21]

decorative and hold religious significance, symbolizing paradise through their intricate designs.

In this case, the Inca textile artistry of Peru (Fig. 4) remains exceptional. Textiles ranked higher than gold and signalled social status through different weaving techniques. These textiles were made from vibrant colours dyed on alpaca and llama wool, with geometric patterns strongly integrated into their cosmology and mythology [22].

In West Africa, kente cloth (Fig. 5) and the admired cloth of Ghana and Nigeria are very symbolic. Kente, a combination of silk and cotton cloth, expresses social status and history. The resist-dyeing method of Adire is recognized for its beautiful patterns on fabrics [24].

Fig. 5 Kente cloth [23]

Fig. 6 Navajo Blanket [25]

Navajo textile productions (Fig. 6) in the American Southwest—such as rugs and blankets—were initially simplistic designs but developed into sophisticated patterns due to cultural interaction with the Spanish and other tribes.

From Indonesia's batik, which UNESCO has recognized for its cultural importance, to the rich tapestry traditions of Scandinavia, the textile practices of each region showcase distinct artistic and cultural stories. In Central Asia, Uzbekistan's suzani textiles and the Matyó embroidery from Hungary in Eastern Europe also reflect deep cultural heritage and skilled craftsmanship. These textiles, celebrated for their spiritual significance, remain highly sought after today. Native American blankets, with their geometric designs and Indian textiles featuring paisley motifs tell stories, share beliefs, and reveal histories unique to their origins. In many cultures, making textiles brings people together, with skills like weaving, embroidery, and dyeing passed on through families to keep the culture alive.

5 Strength of Textile Design: Creativity, Design, and Technology

Textile design showcases creativity through the clever use of colour, pattern, texture, and material. Designers try different fibre weaving techniques and digital tools to make groundbreaking fabrics that push the boundaries of looks and function. Irit Dulman is an artist working on botanical prints and eco-printing, creating pathbreaking strong colour palettes using natural materials [26].

The British designer Zandra Rhodes has become a visionary for her iconic bold colours and hand-printed textiles, drawing ideas from nature and various cultural patterns. In contrast, modern designers like Iris van Herpen blend tech with Fashion. They add 3D printing to textiles, creating futuristic and sculptural clothes. Schiaparelli is another groundbreaking Avant Garde Italian designer popular for witty accessories, surrealist patterns, and loud and striking colour palettes, enabling a more expansive landscape for design and Fashion with its creative couture.

Indeed, considering the issue of microplastics from polyester fabrics and various other pollution alarms from the textile industry, sustainability is no longer a choice for a startup or an established brand. Many designers today are gaining recognition for their innovative contributions to textile sustainability, earning accolades for their dedication to eco-friendly practices and circular design. Sustainable innovations are at the forefront as the fashion and textile industries increasingly focus on sustainability. Designers now utilize eco-friendly materials such as bamboo, recycled polyester, and organic cotton. Additionally, closed-loop recycling systems, biodegradable textiles, and waterless dyeing techniques are being developed to minimize the environmental impact of textile production.

Textile designers are constantly exploring permutations and combinations with wastes of different kinds, namely e-waste, plastics, agro and food processing waste, fungi and bacteria, moulds, weed plants, microorganisms, industrial waste, and many others. Some of the notable examples are Suzanne Lee's "BioCouture," which involves growing biodegradable fabric from microbial cellulose and Aniela Hoitink's MycoTEX, which investigates the use of mycelium to produce biodegradable textiles [27].

Stella McCartney, a trailblazer in sustainable Fashion, has received numerous awards for using organic materials, vegan leather alternatives, waste reduction practices, and maintaining luxury and style. These projects have attracted attention for their potential to lessen the environmental footprint of fabric production. These designers and initiatives highlight the increasing significance of sustainability in textile design as they delve into eco-friendly materials, ethical production methods, and waste reduction to shape the future of Fashion more responsibly [28].

The advent of technology provides new avenues for creativity and efficiency in designing textiles. Computer-aided design tools help designers create patterns and edit them accurately, hence better control of colour, scale, and repetition. CAD also gives greater precision and lessens the opportunities for mistakes in the making process. Furthermore, 3D printing is being investigated to produce intricate, non-traditional textile forms, enabling the creation of innovative structures and textures that surpass the capabilities of conventional methods. Another example is integrating technology into textiles, highlighted by Google's collaboration with Levi's on "Project Jacquard." This initiative weaves conductive threads into denim, connecting the clothing with digital devices. These transformations are reshaping the textile industry.

6 Opportunities: Slow Fashion and Sustainable Fashion

The textile industry is increasingly adopting sustainable design, technology, and management practices, strongly emphasizing waste reduction, supply chain optimization, and durability enhancement. Approaches such as zero-waste patterns, modular designs, and reduced packaging are becoming more popular. A key element of this sustainability movement is community engagement, which encourages collaboration among businesses, local artisans, and marginalized groups to create a fair fashion ecosystem [29].

Supporting local artisans and traditional crafts is essential for preserving cultural heritage and fostering sustainability. These artisans bring unique skills handed down through generations, enriching the textile production process with diversity and meaning. Brands like Alabama Chanin illustrate this by hiring local artisans to produce handmade garments from organic cotton, bolstering the local economy and community. Furthermore, organizations like the World Crafts Council promote traditional crafts, advocating for artisan-made products and raising awareness about the importance of cultural preservation, fair trade, and the economic advantages of supporting local craftsmanship.

New materials and technologies open various opportunities to young researchers and entrepreneurs. Bacterial dyeing presents an innovative and sustainable approach to the environmental challenges associated with traditional dyeing techniques, which typically require significant water usage and involve harmful chemicals. This method utilizes microbes such as Streptomyces coelicolor to generate natural pigments in controlled settings, producing no toxic byproducts. Another interesting experiment was bioengineered spider silk by Bolt Threads, which represents another significant leap forward in sustainable textiles. This lab-produced silk replicates the characteristics of natural spider silk, which is stronger than steel while remaining lightweight and flexible, along with the potential to replace nylon. Shape-memory textiles introduce a cutting-edge concept in clothing design. These fabrics can "remember" and revert to specific shapes when subjected to heat or moisture, enabling garments to adapt to the wearer's body or changing environmental conditions. This versatility makes them particularly suitable for sportswear and adaptive fashion, with researchers actively exploring the integration of shape-memory polymers into textiles.

Self-healing fabrics draw inspiration from nature's remarkable ability to mend wounds, Bio-fabrication using mycelium by MycoWorks, piezoelectric fabrics that generate electricity through mechanical stress like bending or stretching, algae-based textiles, whole garment knitting, 3D weaving and knitting, oleophobic textiles that repel oil and water, and cold plasma technology for textile finishing are some advancements pushing the boundaries of textile design. These innovations make the industry more innovative, sustainable, and increasingly integrated with technology. As these lesser-known technologies develop, they are set to play significant roles in the future of Fashion, opening up new possibilities for creativity, functionality, and environmentally conscious design.

7 Challenges: Artificial Intelligence (AI)

The integration of AI into textile design brings both advantages and challenges. AI boosts creativity by generating unique patterns and designs through sophisticated algorithms, enabling designers to explore fresh ideas and streamline their processes. Various tools make the design process take less than a minute, along with the entire simulation of boards and the final product mockups [30].

AI also facilitates personalized designs by analysing consumer preferences, while predictive analytics assist in forecasting trends, optimizing inventory, and minimizing waste, which supports sustainability [31]. However, the high costs associated with AI technology pose a challenge for small businesses, and poor-quality data might result in less effective designs. Furthermore, issues related to traditional craftsmanship, ethics, and intellectual property rights require careful attention when being overly dependent on computers rather than embracing nativity and creativity.

One significant drawback is that AI often misses the nuanced understanding and emotional depth that human designers bring to their work. For instance, while AI can analyse current trends and create designs based on large datasets, it may not fully grasp the cultural or emotional significance behind a design like a human can. Moreover, AI systems are limited by the data they are trained on; if this data is biased or lacks diversity, the designs produced can reflect those biases. For example, an AI trained primarily in Western fashion trends might overlook or misinterpret styles from other cultures.

8 Conclusion

Recent studies in textile design strongly emphasize sustainability, smart textiles, and innovative materials. A primary focus is on eco-friendly fabrics, including those made from biodegradable materials like mushroom leather, algae-based textiles, and recycled plastics. Researchers are discovering ways to reduce environmental impact through closed-loop production systems that minimize waste and encourage circular Fashion. Smart textiles are another expanding area, where fabrics are embedded with sensors or electronics to respond to stimuli such as temperature, moisture, or light. For example, thermochromic textiles change colour in reaction to body temperature, while wearable technology combines fabrics with health monitoring or communication devices. Biofabrication is also gaining popularity, with researchers investigating using living organisms like bacteria and yeast to produce biodegradable and functional textiles. In the field of digital textile design, 3D printing and digital printing have enabled precise, customizable designs, cutting down on material waste and allowing for intricate patterns that were previously difficult to create. Additionally, research into cultural preservation is blending traditional techniques with modern technology to sustain and innovate ancient textile practices, resulting in a combination of old and new in Fashion. These advancements are paving the way for a future where textiles are more sustainable, interactive, and functional, transforming our understanding of Fashion.

References

1. Gürcüm, B. H. (2017). Conceptual design method and creativity in textile design. *Journal of Textile Engineering and Fashion Technology, 3*(1). https://doi.org/10.15406/jteft.2017.03.00086
2. Singh, J. (2024). Textile design and technology: Past, present, and future. *Redshine Archive, 11*(4). https://doi.org/10.25215/9358099984.27
3. ATIRA. (2023, March 6). *Protective textiles | Applications | Protective clothing compilation.* https://atira.in/protective-textiles-guide/
4. STG. (2024, July 11). *Advancements in 3D printing for fashion and textiles.* Objective3D. https://www.objective3d.com.au/resource/blog/advancements-in-3d-printing-for-fashion-and-textiles/
5. Sporn, S. (2024, February 21). *For Iris van Herpen, fashion's intrepid explorer, anything is possible.* Art Basel. https://www.artbasel.com/news/iris-van-herpen-dutch-fashion-designer-couture-musee-arts-decoratifs-paris-2024?lang=en
6. Manufactur3D Team. (2020, October 14). *How Nike is leveraging 3D printing in the footwear industry.* Manufactur3D Magazine. https://manufactur3dmag.com/how-nike-is-leveraging-3d-printing-in-the-footwear-industry/
7. Saunders, S. (2022, August 10). *Ganit Goldstein uses Stratasys 3D printing to create gorgeous GnoMon fashion collection.* 3DPrint.com: The Voice of 3D Printing. https://3dprint.com/293120/ganit-goldstein-uses-stratasys-3d-printing-to-create-gorgeous-gnomon-fashion-collection/
8. Dekel, S. (2024, November 7). *Danit Peleg on 3D printed fashion design with zero waste.* Culture Treasures. https://culturetreasures.com/danit-peleg/

9. Li, Y. (2022). Application and development of computer-aided design methods in clothing design under the environment of internet of things. *Mobile Information Systems, 2022*, 1–9. https://doi.org/10.1155/2022/3405540
10. Becher, B. (2024, November 25). *What are smart textiles?* Built In. https://builtin.com/articles/smart-textiles
11. Júnior, H. L. O., Neves, R. M., Monticeli, F. M., & Agnol, L. D. (2022). Smart fabric textiles: Recent advances and challenges. *Text, 2*(4), 582–605. https://doi.org/10.3390/textiles2040034
12. Graziola, E. (2024, July 31). *New textile revolution: Smart textiles and wearable electronics*. Datatex. https://datatex.com/new-textile-revolution-smart-textiles-and-wearable-electronics/
13. Abrahart, E. N., & Whewell, C. S. (2024, November 8). *Textile*. Encyclopedia Britannica. https://www.britannica.com/topic/textile
14. Textile Value Chain. (2024, June 25). *The history of textiles: From ancient times to modern day*. Textile Value Chain: Textile Magazine, Textile News, Apparel News, Fashion News. https://textilevaluechain.in/news-insights/the-history-of-textiles-from-ancient-times-to-modern-day
15. Taylor, G., & Quirke, S. (n.d.). *Textile*. https://www.ucl.ac.uk/museums-static/digitalegypt/textil/tools.html
16. Kenoyer, J. M. (2004). Ancient textiles of the Indus Valley region. In N. Bilgrami (Ed.), *Tana Bana: The woven soul of Pakistan* (pp. 18–31). Koel Publications.
17. *File:Kosode with yuzen dyeing inside fan and snowflake shapes, 1 of 2, Edo period, 1700s AD, chirimen crepe—Tokyo National Museum—Tokyo, Japan—DSC09592.jpg—Wikimedia Commons*. (2018, April 15). https://commons.wikimedia.org/wiki/File:Kosode_with_yuzen_dyeing_inside_fan_and_snowflake_shapes,_1_of_2,_Edo_period,_1700s_AD,_chirimen_crepe_-_Tokyo_National_Museum_-_Tokyo,_Japan_-_DSC09592.jpg
18. Gupta, M. (2021, January 12). *The fascinating textile crafts of Japan*. Marasim—Sustainable Textile, Craftsmanship, Luxury. https://blog.marasim.co/the-fascinating-textile-crafts-of-japan/
19. *File:Turkey, Ushak, Ottoman period—Classical Turkish Carpet with the Lotto Pattern—2015.61—Cleveland Museum of Art.jpg—Wikimedia Commons*. (n.d.). https://commons.wikimedia.org/wiki/File:Turkey,_Ushak,_Ottoman_period_-_Classical_Turkish_Carpet_with_the_Lotto_Pattern_-_2015.61_-_Cleveland_Museum_of_Art.jpg
20. *File:Persian Silk Brocade—Shah Abbasi Flower—Seyyed Hossein Mozhgani—1974.jpg—Wikimedia Commons*. (2011, February 5). https://commons.wikimedia.org/wiki/File:Persian_Silk_Brocade_-_Shah_Abbasi_Flower_-_Seyyed_Hossein_Mozhgani_-_1974.jpg
21. *Figure File:Tupa-inca-tunic.png—Wikimedia Commons*. (n.d.). https://commons.wikimedia.org/wiki/File:Tupa-inca-tunic.png
22. Cartwright, M. (2015, February 1). *Inca textiles*. World History Encyclopedia. https://www.worldhistory.org/article/791/inca-textiles/
23. *Figure File:Kente cloth is colorful.jpg—Wikimedia Commons*. (2024, March 5). https://commons.wikimedia.org/wiki/File:Kente_cloth_is_colorful.jpg
24. Fening, K. O. (2006). *History of kente cloth and its value addition through design integration with African wild silk for export market in Ghana*.
25. *File:America, Native North American, Southwest, Navajo, Post-Contact, Transition—Blanket-Sarape (banded style)—1921.948—Cleveland Museum of Art.jpg—Wikimedia Commons*. (n.d.). https://commons.wikimedia.org/wiki/File:America,_Native_North_American,_Southwest,_Navajo,_Post-Contact,_Transition_-_Blanket-_Sarape_%28banded_style%29_-_1921.948_-_Cleveland_Museum_of_Art.jpg
26. Shukin, N. (2023, July 17). *Eco print—Artistic way to bring nature into your home*. Iritdulman. https://iritdulman.com/eco-print/
27. Llonch, L. (2022, August 5). *BIOCOUTURE: Garments with living bacteria*. Knowledge Hub. https://knowledge-hub.circle-economy.com/wctd/article/18396?n=BIOCOUTURE-Garments-with-living-bacteria
28. Bajekal, N. (2023, August 9). Stella McCartney on wanting to make fashion more sustainable. *Time*. https://time.com/6302562/stella-mccartney-sustainability-interview-lvmh/

29. Siliņa, L., Dāboliņa, I., & Lapkovska, E. (2024). Sustainable textile industry—Wishful thinking or the new norm: A review. *Journal of Engineered Fibers and Fabrics, 19*. https://doi.org/10.1177/15589250231220359
30. Kulkarni, K. (2024, July 9). *The evolutionary impact of artificial intelligence on the textile industry*. Fibre2Fashion. https://www.fibre2fashion.com/industry-article/10074/the-evolutionary-impact-of-artificial-intelligence-on-the-textile-industry
31. Kapoor, S. (2024, January 12). *Power of artificial intelligence in the textile industry*. GreyB. https://www.greyb.com/blog/artificial-intelligence-in-textile-industry/

Textile Design in Interior Spaces

Oinam Roselyn Devi, Arpita Grover, and Bhawana Chanana

1 Introduction

Interior design became a recognized profession as society and economics developed. In the United States, interior decoration and interior design are often considered synonymous. In contrast, in India, the tradition of decorating grand residences called havelis with intricate murals dates to the medieval era. Between the seventeenth and nineteenth centuries, interior decoration became a key concern for homemakers, reflecting their aesthetic preferences. Meanwhile, architects began to use interior design to collaborate with craftsmen on building projects. The mid- to late nineteenth century saw a significant expansion of interior design, driven by the growth of space and the rise of the middle class in industrialized nations. This development allowed people to showcase their wealth and social status through their living environments [1].

Interior design and décor can be confusing to many people, which leads to serious issues. Interior design is the process of creating a whole living area from the ground up, whereas interior decoration is the art of adding beauty to an already-existing living area through upholstery. Interior design is not as easy as one may think; it is a science constrained by its own elements and principles [2]. Interior designers typically follow the Book of Rules, which lists seven components and seven principles of interior design. With specific volume and surface change techniques, interior design improves the experience of sculpting the interior space. Interior design addresses environmental psychology, architecture, and advancements in production design in addition to traditional ornamentation. The process of

O. R. Devi (✉) · A. Grover · B. Chanana
Amity School of Fashion Technology, Amity University Mumbai,
Mumbai, Maharashtra, India
e-mail: ordevi@mum.amity.edu; agrover@mum.amity.edu; bchanana@mum.amity.edu

Sadhna et al. (eds.), *The Art and Craft of Modern Textile Design*, Sustainable Textiles: Production, Processing, Manufacturing & Chemistry,
https://doi.org/10.1007/978-3-031-86797-2_9

constructing and arranging features inside several types of constructions is known as interior design. These components could be furniture, ornamental pieces, surface treatments, and color palettes. This procedure is applied not just in homes but also in workplaces, shops, and even public spaces like dining establishments and airports [3].

Among the numerous materials used in interior design, textiles are noteworthy. A room that appears harsh or stark can be made to appear warm, cozy, and humane by incorporating them into it. Their vast array of hues, textures, and other characteristics give a static area a feeling of softness, curve, and flexibility. They offer an infinite number of design possibilities as well. According to a prior post titled "Recognizing textiles as a significant contributing factor to interior design as opposed to conventional understanding," textiles occupy between 60% and 70% of interior spaces in residential as well as business environments. It could be because of their innate decorative potential that textiles are usually selected solely based on their aesthetic elements. Residential (often referred to as domestic) and commercial (commonly referred to as contract) are the two main markets for interior textiles. Both areas make interior textiles. While contract textiles are found in public or commercial settings, domestic textiles are found in homes or private interiors [4]. The two main categories of interior textiles that are typically mentioned are home textiles and furnishing fabrics. Upholstery fabrics, wall coverings, soft floor coverings, window furnishings (blinds, drapes, and curtains), and accessories like throws and cushions are examples of furnishing fabrics. Apart from furnishing materials, all textiles used in interior spaces of homes are considered household textiles. For instance, napkins, tablecloths, blankets, towels, and beds. Products that are made for care and hospitality facilities are known as "institutional textiles" because they must meet strict performance and safety standards [5]. The main purpose of furnishings and indoor textiles is to provide practical and functional purposes, like covering and shielding surfaces. Conversely, they are also important for décor, comfort, and well-being, reflecting messages about personal comfort, choice, and lifestyle. These things have long been a part of everyday life, from woven carpets to printed upholstery fabrics and ancient woven textiles. Advanced textile technology and synthetic fibers have replaced traditional natural materials in recent years. These high-quality materials are now widely employed to make futuristic, innovative fabrics for contemporary architecture and interior design [5].

Nomadic Bedouin communities have historically constructed enormous woven tents and all the inner fabrics required for their nomadic way of life in many regions of the world. Animal hair was used to make a wide range of products, including camel decorations, floor coverings, bedding, storage bags, tent dividers, and cushion covers. In addition to providing shelter from harsh weather and intertribal conflict, these fabrics also created a cozy and practical living space made completely of textiles. Textiles are now used extensively in modern architecture as well as in the functional and decorative aspects of our houses. Bedding, carpets, and curtains are examples of items that not only offer warmth but also express our sense of wealth and lifestyle choices. Certain demands, including thermal insulation or light-softening transparent membranes, are catered for by specialty fabrics. In public areas and business buildings, textiles improve comfort and atmosphere while also enhancing brand identification and status. As technological and specialty textiles, such as cutting-edge polyester-based textiles, continue to evolve, designers and engineers are able to develop innovative forms of interior art and architectural solutions [6]. The history of textile printing is extensive and traces back to the origin of human civilization. The ancient wall paintings of the Ajanta Caves in India, which date to the seventh century BC, provide evidence of early textile printing processes. These paintings demonstrate intricate dyeing and patterning techniques, and textiles are frequently used by royalty and the upper classes to signify high status in society. Techniques for resist dyeing and block printing became widely used throughout Asia and the Mediterranean after Alexander's invasion of India in 327 BC. This led to the popularity of Indian textiles in Europe, where they were reproduced commercially. The design and manufacture of domestic textiles advanced because of discussions with the British East India Company in the seventeenth century and ongoing textile exchanges with Arab traders in the second century AD. In the early eighteenth century, during the Rococo period in Europe, intricate interior prints gained popularity. Francis Nixon developed the copper-plated printing method in Ireland, which resulted in major advancements like better print quality, larger repetition motifs, and faster printing. By the 1930s, textile printing was introduced to the West and became mechanized after World War II. The design and accessibility of items like upholstery, curtains, floor coverings, bedding, and tableware were revolutionized because of the mass production of standard-controlled furnishing materials [7].

Furnishings and interior textile items were able to quickly adjust to shifting styles and modern designs during the 1960s and 1970s because of the fast-developing textile industry, which was defined by high-speed, flexible, and economical production methods. High-end domestic and contract markets employed natural materials for upholstery textiles, tapestries, curtains, rugs, dinnerware, towels, and mattresses as consumers valued their ease and wellness benefits. Conversely, more recently, the emergence of reasonably priced technical and synthetic textiles has enhanced their appeal and caused many natural materials to be replaced by them. Nowadays, man-made fibers are used extensively in upholstery items, such as padding, filler, and barrier materials for furniture and seats. These products must adhere to strict health and safety standards, including being fire and flame resistant, and are used in houses, cars, airplanes, and boats. Interior textiles are made for their functional qualities, such as light fastness and abrasion resistance, as well as their visual appeal. Traditionally, traditional looms were used to hand-weave wool and silk into carpets. Modern, advanced companies like Wilton and Axminster Carpets now create luxurious carpets and rugs in a wide range of hues and patterns to improve interior design. Synthetic fibers and natural materials are combined to maintain high standards of quality, safety, and health while keeping these items reasonably priced [7].

2 Home Decor Trends

2.1 Innovative Home Décor Design Using Old Wood

Ash, white oak, and black walnut are only a few of the many wood species that exist because of the differences in growth location and growth speed. Wood has rather distinct patterns and qualities as well. Wood has become the most popular resource in the world of interior decoration because of its unique creative impression, superior durability, and ease of processing and use. It can be used on a variety of indoor surfaces, including walls, ceilings, doors, windows, floors, furniture, tiny installation materials, and crafts related to furnishing. It performs exceptionally well, beyond that of other materials. Waste wood design work reflects a taste of the past and the good old days in contrast to design work created with new wood. The use of scrap wood manufacturing for creative home décor is therefore becoming increasingly popular among consumers and is becoming a fundamental component of contemporary interior design [8]. Reusable and recyclable wood supplies are widely available nowadays for furniture and building. A comprehensive plan should be in place to deal with the ancient woodlands and recycle these resources. Old wood from the architecture and home furnishings markets can be recycled and repurposed to create wooden handicrafts and innovative furniture that can be used to adorn a home, restaurant, theme bar, or tea house in the traditional style [9].

Chen and Peng use discarded wood to make a variety of embellishments for home furnishings. These authors employed a variety of design techniques while creating home furnishings, including the molding technique for wood assembly, the cutting and recombining technique, the surface coating and varnishing technique, and the wood carving technique [10].

2.2 Wood Waste as a Source for Pigments for Water-Borne Wood Paints

Calovi and Rossi used wood waste as a source of pigments for water-borne wood paints. The authors assess how an unusual black pigment derived from wood waste affects the functionality of bio-based wood paint [11]. This industrial pigment, called *BioBlack TX*, was manufactured and supplied by Nature Coatings Inc. The company uses only certified Forest Stewardship Council (FSC) wood waste from the paper, lumber, furniture and flooring industries. The wood is burned in oxygen-free chambers, which stops the wood from burning and, consequently, from generating CO_2. As a result, wood effectively becomes carbonized and absorbs the carbon that would have otherwise been released during the burning process. The pigment can therefore be described as carbon-negative. The facility's components are powered by the heat and biogas produced during the production process, which also yields the patented pigment. Additionally, the pigment is 100% biobased and provided as an aqueous dispersion. To coat or print on surfaces including leather, paper, textiles, and film, the product was produced as an aqueous dispersion of black pigment.

2.3 *Low-VOC Paints*

Furniture treated with paints, varnishes, waxes, and solvents frequently releases volatile organic compounds (VOCs) [12–15]. The quality of indoor air is impacted by the substantial VOC emissions produced by furniture constructed of solid wood and wood composites. VOCs may be harmful to children's health because of their ongoing physical and cognitive development, which makes them more vulnerable. Depending on the amount and duration of exposure, VOCs might have different effects. The central nervous and respiratory systems are the main organs affected by acute, brief exposure, which can cause symptoms like headaches, dizziness, and irritation of the eyes, nose, and throat. Long-term, chronic exposure may have more detrimental effects on health.

On the other hand, long-term exposure to VOCs poses more serious health hazards, impacting the respiratory, hematological, immunological, and central neurological systems. Serious illnesses including immunodeficiency, which affects blood chemistry and raises the risk of leukemia, can result from this extended exposure. Additionally, it can lead to respiratory disorders including asthma, peripheral neuropathy, memory loss, balance problems, sluggish reaction times, and concentration problems [16, 17]. VOCs can penetrate the blood-brain barrier since they are lipid-soluble, which could result in serious neurological health issues.

"A Guide to Indoor Air Quality," published by the US Environmental Protection Agency (EPA), emphasizes that indoor air pollution can frequently present more risks than outdoor air pollution. The American Society of Heating, Refrigerating, and Air Conditioning Engineers (ASHRAE) defines indoor air quality as maintaining contaminants' concentrations below specified limits [18]. In a comparable way, the 2017 report "Revision of European Union Ecolabel Criteria for Furniture Products" emphasized how seriously VOCs released from furniture affect indoor air quality. This report suggested a flexible approach to lowering emissions, eliminating the requirement for costly tests while recommending the use of low VOC coatings and assessing VOC emissions from finished items [19]. Toluene, formaldehyde, and benzene derivatives are examples of typical volatile organic compounds (VOCs).

2.4 Nanomaterials Applications in "Green" Functional Coating

Wood is highly valued in the furniture and building industries for its renewability, environmental friendliness, and high strength-to-density ratio. However, it is vulnerable to damage from oxygen, light, and water and is also prone to microbial and insect attacks and fire hazards. To safeguard wood, an appropriate paint coating is essential. Although solvent-based paints, like polyurethane (PU) and cellulose-based alternatives, have long been used for wood finishing because of their consistent performance, they contain hazardous components like heavy metals (such as copper, chromium, lead, and arsenic) and volatile organic compounds (VOCs), which greatly endanger human health and the environment. As regulations become more stringent, there is a pressing need for high-performance, water-based environmental paints. These water-based alternatives generally have lower levels of VOCs and heavy metals, but they often face challenges such as longer drying times, limited functionality, and restricted chemical tunability [20–23].

To address the issues associated with traditional wood coatings, a new generation of "environmental paints" has been developed using organosilica nanosols. These paints incorporate desirable properties through chemical modifications of the nanosol and the addition of nanomaterials and functional additives. Organosilica is a unique kind of chemical that combines the strength and endurance of inorganic coatings with the adaptability of conventional organic and polymeric materials. It has a hydrophilic, hydrolyzable silane head and hydrophobic arms. This allows organosilica to work well with organic and aqueous substances. Therefore, using organosilica to produce high-quality, multipurpose environmental paints is more economical and efficient than using typical polyurethane (PU) and cellulose-based paints, which often require extra additives or processes. Organosilica nanosol-based environmental paints have already shown excellent results in preserving wood [24].

2.5 Biophilic Interior Environments

The primary objective of biophilic design is to improve human well-being by incorporating natural systems and processes into the constructed environment. Drawing from the Biophilia hypothesis, it proposes that people are inherently linked to nature and that engagement with natural settings is essential to our well-being [25]. Nevertheless, urbanization, modern architecture, and shifting lifestyles frequently result in a lack of engagement between modern communities and nature [26]. By integrating natural elements and systems into interior and architectural environments, biophilic design seeks to close this gap, restoring people's connection to nature and enhancing their general quality of life [27].

Biophilic design has been shown to improve human health and well-being in more than 30 years of research. According to James et al., chronic pain, migraines, and depression are among the leading causes of disability worldwide. People with chronic illnesses frequently have severe symptoms and a lower quality of life [28]. Pharmaceutical medications and non-pharmacological therapies like behavioral and physical therapy are used to treat chronic disorders, but neurological diseases are becoming more common worldwide [29]. The causes and symptoms of chronic diseases are made worse by a sedentary lifestyle; individuals in the US, Canada, and Germany spend an average of 15–17 h indoors each day [30, 31]. Physical activity was further reduced by the COVID-19 pandemic's "stay-at-home" limitations, which had a detrimental effect on mental health [32]. According to the World Health Organization, "the ability of individuals, families, and communities to promote health, prevent disease, maintain health, and cope with illness with or without healthcare support" modification of lifestyles and self-care strategies can enhance health outcomes [33–35]. These include restful sleep, mindfulness, music, exercise, a balanced diet, aromatherapy, stress reduction, and spending time in nature.

The advantages of integrating nature into interior spaces are highlighted by recent literature studies on biophilic design and healthy home settings. While Huntsman and Bulaj acknowledged the benefits of biophilic interior design in healthcare facilities, they also highlighted that there hasn't been as much focus on supporting the health of people with chronic illnesses in their homes [36].

Considering the COVID-19 pandemic, which had people staying indoors to avoid infection, home furnishings that support mental and physical health are especially crucial. The study examined the possible health advantages of designing residential rooms with natural elements, such as light therapy, areas for relaxation and meditation, music, exercise, aromatherapy, and improved sleeping conditions. The goal of combining biophilic design with therapeutic environments is to improve patient care and reduce the symptoms of diseases like depression. In the same way, Zhong et al.'s study from 2022 highlighted that although "nature" and biophilic design have received a lot of attention in the field of architecture, there are still a lot of unresolved concerns and conflicts about how nature is viewed, handled, and incorporated into the design [37]. The study investigated biophilic design as a theoretical framework for understanding the importance of nature in architectural engineering through an overview of the literature. It examined the advantages of biophilic design in fostering sustainability, productivity, well-being, and health by contrasting its key frameworks and components. The results highlighted that biophilic design is more than just integrating plants into constructed areas; it also entails enhancing the implementation of biophilic design in general and integrating more types of nature into the built environment.

3 Textile Design

Conventional textiles are often composed of materials that are rich in chemical components produced throughout the pre-finishing, finishing, dyeing, and weaving stages of the textile's production process. It is very dangerous to utilize such chemical components in textiles in interior spaces because they can cause a lot of problems and allergies. As a result, as textiles comprise most of the interior space, interior designers should switch from utilizing standard textiles to eco-friendly ones to help create a sustainable and healthy interior environment. Additionally, it is recommended that textile suppliers be informed of this so that they can set aside areas of their stores for green or eco-friendly textiles, providing customers with the choice to choose eco-friendly fabrics over conventional ones to prevent illnesses and allergies and lessen their negative effects on the environment. Some fabrics, especially those that have undergone chemical processing to give them specialized properties, are utilized in interior areas and contribute to indoor air pollution. This degrades the quality of the air inside, either directly or indirectly. Hazardous materials can be released out of textiles in several ways throughout use, laundering, and eventual disposal as waste. Consequently, the chemicals in textiles will be exposed to both people and the indoor environment.

3.1 Eco-Friendly Textiles Used in Interior Spaces: Organic Textiles/Eco-Friendly

3.1.1 Linen

Due to its low need for artificial fertilizers, pesticides, and water, flax is an environmentally friendly plant. The whole plant is used, which results in very little energy consumption and very little or no waste while producing linen from flax. Tablecloths, draperies, and upholstery are used in spaces [38].

3.1.2 Bamboo

Bamboo fabric is renewable and biodegradable because it comes from the bamboo plant. The fiber used to make textiles from bamboo has no potentially harmful elements because the plant grows quite quickly without requiring a lot of water or chemical fertilizers. The primary factor that makes it an environmentally beneficial fiber is its renewable nature. The fabric has inherent antibacterial properties and breathes. Upholstery, wallpaper, drapes and curtains are employed in interior spaces due to their insulating properties [38].

3.1.3 Organic Wool

Wool is a fire-resistant, renewable material that doesn't need any chemical additives. Increasingly, wool produced using sustainable farming practices is becoming available. Wool is a very useful fabric that has several important characteristics that make it quite attractive. One of wool's primary benefits is its capacity to efficiently absorb moisture by pushing moisture to the core of its fibers. Wool is used in making upholsteries, blankets, rugs, and carpets [38].

3.1.4 Jute

One of the greenest fabrics produced worldwide is jute. This is due to its full biodegradability. Additionally, it releases oxygen and absorbs carbon dioxide. It can improve the fertility of the soil it grows in, doesn't release microfibers, and doesn't pollute rivers. It can be grown without the use of pesticides or fertilizers. It is not only sustainable but also environmentally friendly. Jute is also used in interior spaces such as curtains, upholsteries, carpets, rugs, and outdoor furniture [38].

3.1.5 Soy Silk

An eco-friendly, fully biodegradable textile was made from tofu production waste. The long, continuous threads of liquid soy protein can be crushed and spun much like any other fiber. The high protein content of soy eliminates the need for artificial coloring, and the fabric responds especially well to natural colors. Curtains, cushions, and upholstery are among the interior design applications for soy silk [38].

3.1.6 Lumpur

Manufactured from the trimmings of white pine trees, this biodegradable fabric "offers the softness of silk, the touch of cashmere, and the lightness of linen. "Lumpur" is better than other cellulose fibers because it is softer, can absorb and release moisture, and can resist a greater variety of temperatures. The latter feature enables Lumpur to maintain a warm atmosphere in the winter and a cold climate in the summer. Interior areas are furnished with drapes, upholstered furniture, and rugs [38].

3.1.7 Fish Skin Leather

Fish skin leftovers from the food business, such as cod, carp, sturgeon, catfish, salmon, perch, and wolfish, are used to make fish skin leather. Leather can be used to make belts, handbags, clothes, shoes, accessories, furniture, and interior design pieces [38].

3.1.8 Lotus Stem

The fibers from lotus stems that have been extracted and hand-spun into yarn are used to make this cloth. Saints wore it in the past, and it is regarded as one of the most durable and sacred textiles in the world. These handloom textiles, which are mostly handwoven, resemble silk and linen in both appearance and texture. These are silky, wrinkle-free, stain-resistant, and waterproof [38].

3.1.9 Banana Fiber

The stems of the banana plant are used to make banana fabric. The lengthy strands provide silky threads that are used to make rugs and other home textiles [38].

3.1.10 Nettle Fabric

This is made from the stinging nettle plant's widely accessible and popular stem in Uttarakhand. Due to its low water use and ability to be cultivated and then replanted in the field to restore depleted nutrients, the plant is considered sustainable. Nettle plant textiles also garner attention due to their environmental friendliness and lack of pesticide use. In addition to being finer, more elastic, and stronger than linen, nettle fabric has a velvety, luxurious appearance. Nettle plants are used for blue dye, cow feed, upholstery stuffing, nets, bags, sacks, coats, mats, vests, shawls, and strings [38].

3.2 Textile Farming

"Textile Farming" investigates the design of a "textile interior ecosystem," looking at things like growing crops, composting, and fermenting in interior spaces. It speculates on how these biological principles might be applied to textile design and how this might alter how we create interior spaces, coexist with them, and perceive them as a community of living systems. The life cycle of biological agents, such as bacteria and plants, is given a biological viewpoint by being incorporated into indoor textiles. Here, the textile's functionality is expanded by additionally acting as a growth template, a substrate for biological agents, and a mediator for all parties participating in the "textile interior ecosystem." To create hybrid materials for interior design, the ongoing Textile Farming research program combines textile structures with seeds to investigate alternative plant organization techniques [39].

3.3 Designing and Constructing Various Home Decor

In the modern day, designing and producing a variety of home décor goods by fusing diverse batik and natural dyes offers a creative and approachable way to support small business owners. By combining interesting patterns with batik and natural dye on cambric cotton cloth, Kaur created wall frames that fused history and contemporary. The dyes were extracted from natural elements such as marigold flowers, turmeric, supari, tea, and onion peels. The wall frames were made from cardboard, glass, and a wooden frame [40].

Similarly, five designs were created by Babel and Sachihar using Sanjhi motifs as inspiration [41]. Using Coral Draw software, an extensive sketch of the motif was made. The location of the finished product was extensively considered, and the best placemat arrangement was chosen for product development. Housewives ranked the designs, and five cushions with Sanjhi motifs were made, as illustrated in Fig. 1.

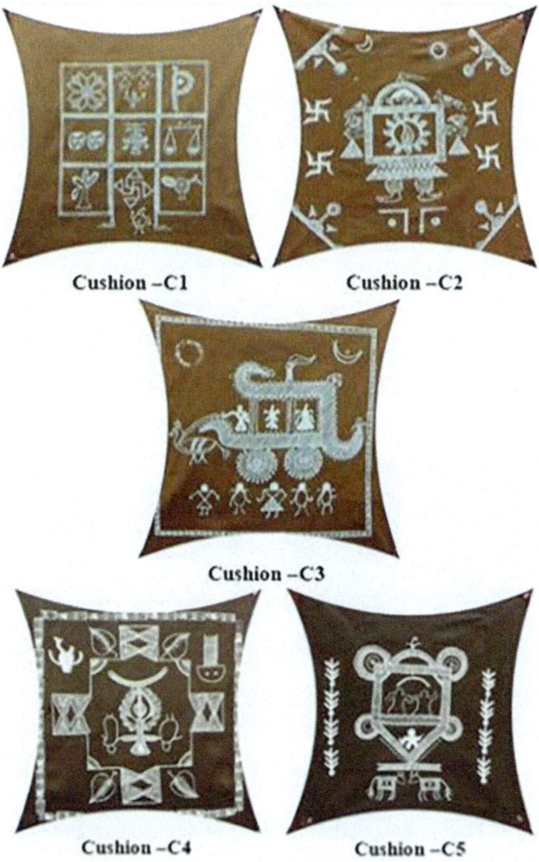

Fig. 1 Babel and Sachihar created five cushions using Sanjhi motifs as inspiration [41]

The cost of labor and the cost of the raw materials needed to make the prepared cushions were added up to determine their final cost.

Generations of artisans have masterfully produced Madhubani paintings, a famed art form from Bihar known for its brightness and meaning. These paintings are a notable example of visual folk art and successfully depict a variety of tribal life elements. Alongside Madhubani paintings, there are other vivid folk-art forms closely tied to the daily life of Maithili rural communities, such as crafts made from natural resources like bamboo and herbal ingredients, as well as arts like siki-mauni, sujni, sitalpati, godhna, and sculptures of mythological figures like sama-chakeva. Gupta and Gangwar developed textile products, including tablecloths, cushion covers, and folders, inspired by Madhubani paintings [42]. France developed six different types of home decor, from post-consumer waste by upcycling and transformable methods. The products include bed-lamp cum holders, hoop wall décor, cushions, magazine holder cum table runners, wall hanging and drum pendant lamps, as shown in Fig. 2 [43].

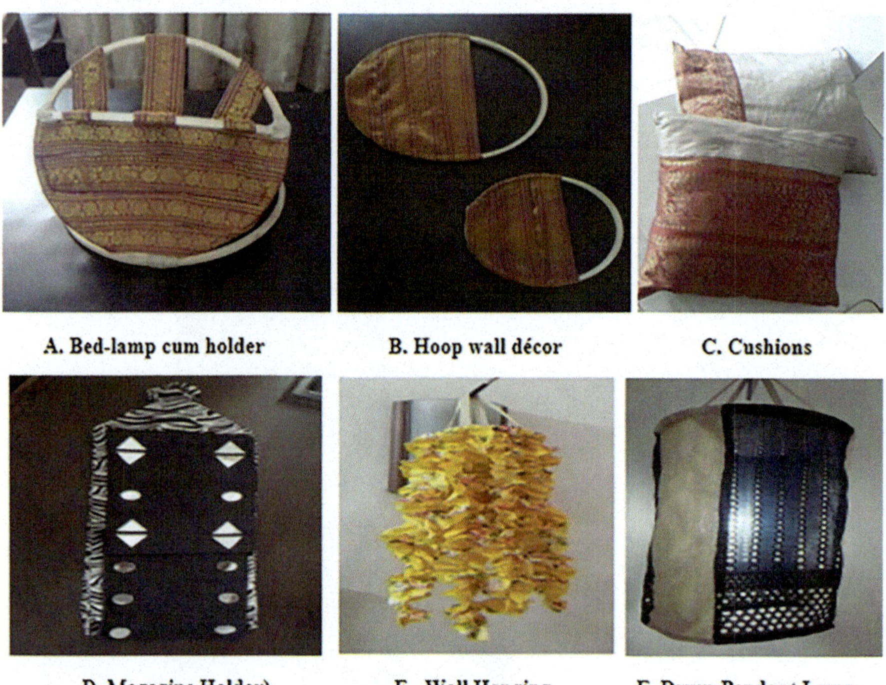

Fig. 2 Designing and development of home décor and home textiles using post-consumer textile waste [43]

3.4 Utilization of Textile Waste into Interior Space

Textile waste and agricultural byproducts from the extraction of unconventional fibers can be utilized in the production of floor and roof tiles. Sisodial and Parmar made tiles with both natural and synthetic resins utilizing three different kinds of materials: waste from denim fabric, cornhusk fiber, and rice straw fiber, as shown in Fig. 3 [44]. These tiles were made to withstand water and fire. These tiles' mechanical and chemical characteristics were evaluated using the IS 13630 standards. The findings demonstrated that the tiles made with synthetic resin and textile and agricultural waste fulfilled Indian Standard standards. Tiles manufactured from denim waste, whether using natural or synthetic glue, showed a higher tensile strength of 18.89 MPa, according to tensile test findings, than tiles made from rice straw (7.54 MPa) and cornhusk (9.91 MPa). Nevertheless, the tiles manufactured from agro-waste and denim waste using natural resins did not meet the standards for some qualities, such as shock resistance, heat stability, and chemical resistance.

In a study by Jamshaid et al., sustainable composite brick samples were made by combining waste synthetic knitted fabric with white cement (WC), acrylic glue (GL), and unsaturated polyester resin (UPR) [45]. Brick prototypes were created during the testing phase by combining the waste fabrics with hydraulic and synthetic binders. Figure 4 shows the schematic illustration for making textile-waste-based bricks. These samples' mechanical and thermal characteristics were examined and contrasted with those of conventional clay bricks. The GL samples had the lowest mechanical strength, while the UPR samples had the highest mechanical strength. The mechanical characteristics of the WC samples were mild. With tensile, bending, and compressive strengths of 0.111, 0.134, and 3.114 MPa, respectively, the UPR samples showed the highest values. The GL samples demonstrated the lowest mechanical performance, with values of 0.0162, 0.0492, and 1.542 MPa, respectively, whereas the WC samples had tensile, bending, and compressive strengths of 0.064, 0.106, and 2.670 MPa, respectively. With a minimum

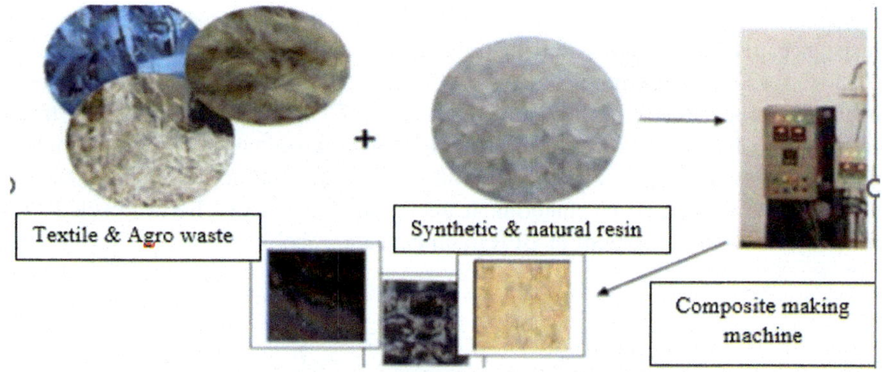

Fig. 3 Process of making tiles from textile and agro wastes with the combination of natural and synthetic resins [44]

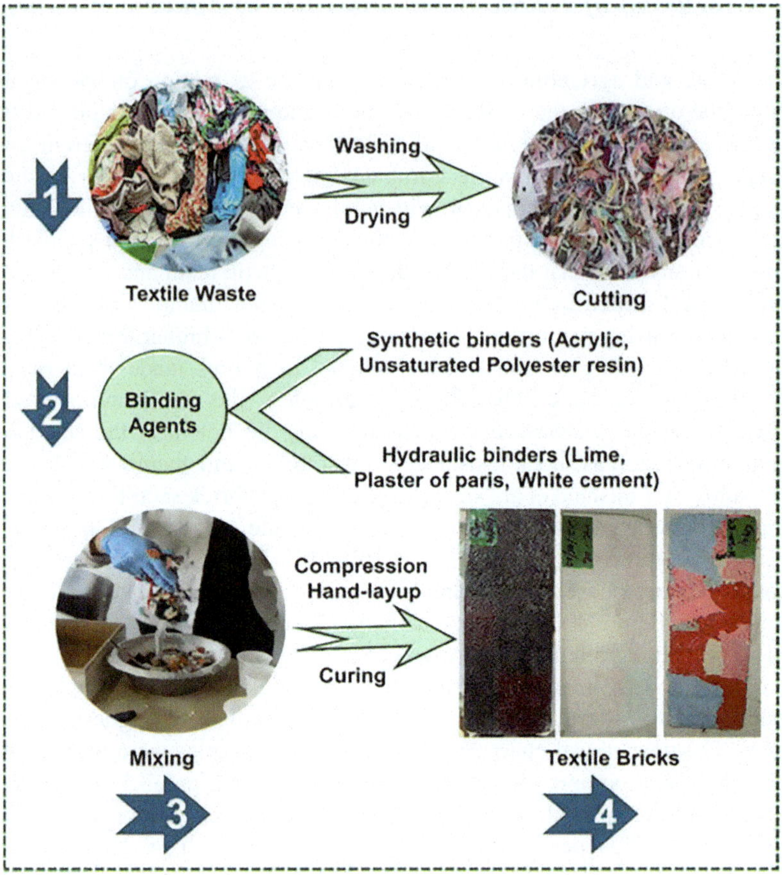

Fig. 4 Schematic diagram for making bricks from textile waste [45]

temperature rise of 54.3 °C after 15 min, WC samples demonstrated remarkable resistance to heat transmission in terms of thermal conductivity, as opposed to 57.3 °C for GL-based samples and 58.1 °C for UPR. UPR demonstrated superior heat resistance in terms of thermal insulation, but GL permitted quicker heat transfer for up to 10 min of heating. To promote more environmentally friendly interior design, this study investigates the possibilities of a circular textile system that uses textile waste as a building material. It also shows how well short, recycled PET fibers work as reinforcement in UPR composites.

4 Custom Textile Solutions for Homes

In the realm of interior design, textiles are more than just functional elements; they are transformative tools that infuse spaces with warmth, character, and style [46]. "Custom Textile Solutions for Homes" explores the integral role of textile design in shaping interior spaces, offering insights into how personalized fabric choices can elevate the aesthetic and functional qualities of a home. Custom textiles, from carpets and cushions to draperies and upholsteries, can be carefully crafted to satisfy requirements, express personal preferences, and create unified, comfortable spaces. Readers will learn how to create rooms that are not only aesthetically pleasing but also customized to meet the specific needs and preferences of the people who live there by comprehending the subtle aspects of textile design.

More and more homeowners are looking for goods that suit their individual preferences and tastes. Customized textile solutions that meet specific needs can now be produced through recent developments in textile manufacturing technology. Three types of design techniques can be distinguished for personalized and customized home textiles: standardized individualization, particular adaption, and special design. Special design entails starting from scratch and producing completely original textile solutions that are tailored to the homeowner's needs and vision. Specific adaptation is the process of altering preexisting designs to satisfy requirements or tastes while striking a balance between originality and effectiveness. Standardized individualization makes customization more affordable and accessible by allowing for customizing within a structured framework using predetermined options and configurations. Examples of tailored textile applications in diverse home environments are used to analyze the features of these design processes. The conversation examines the ways in which early customer interaction and the preparation of client specifications impact the design process for various degrees of customization made feasible by innovative textile production techniques [47].

To improve their goods and give consumers the impression that they are buying something genuinely unique, luxury brands have been offering more and more possibilities to personalize and customize their items. Nevertheless, these personalization choices frequently only offer different takes on preexisting themes. To give the appearance of uniqueness, component parts from an established product line are manufactured and sold as a part of an already-existing category. Although providing personalized products alters the way customers perceive their purchases, giving them the impression that they are getting something unique, the reality frequently falls short. Customization helps luxury brands expand their product lines and increase the perceived value of their offerings [48].

AI-driven personalization is revolutionizing the way luxury firms interact with their consumers as customization and personalization gain popularity in the market. Generative AI can produce customized marketing content that appeals to specific clientele. AI is being used by luxury brands to improve consumer interactions and offer customized shopping experiences. Chatbots and virtual assistants driven by AI enhance customer service by providing prompt, tailored responses, which is particularly beneficial for wealthy customers who want excellent services. Likewise, chatbots and virtual assistants driven by AI provide continually customer service with prompt, tailored responses. The luxury landscape is being redefined by the fusion of cutting-edge technologies with classic premium workmanship, striking a balance between innovation, authenticity, and personalization [49].

5 From Massive Customization to Mass Customization

The two methods of customizing textile solutions for homes—"Massive Customization" and "Mass Customization"—offer varying degrees of efficiency and customization.

5.1 *Massive Customization in Home Textiles*

Massive customization involves producing incredibly customized textile solutions that are suited to each homeowner's particular requirements and tastes. It emphasizes personalized designs and particular client needs, providing a great deal of customization. Usually requiring handcrafted or made-to-order textiles, this method is more costly and time-consuming. Bespoke upholstery, especially patterned rugs created according to specific measurements, and drapes with materials and designs selected by the homeowner to match unusual window sizes are a few examples of extensive customization.

5.2 *Mass Customization in Home Textiles*

The cost-effectiveness of mass production and the versatility of customized home textile solutions are combined in mass customization. Within a uniform framework, it enables homeowners to personalize their selections [50]. To fit their decor, homeowners can choose from a variety of pre-designed slipcovers offered by a home textile firm, for instance, and make minor adjustments to the fabric and color. This strategy uses modular designs that give clients a variety of predetermined alternatives while striking a balance between economies of scale and customization. Massive customization is slower and more expensive than mass customization. Examples include adjustable window curtains, modular furniture slipcovers, and bed linens that may be customized in a range of colors and patterns. Table 1 shows the key differences between Massive Customization and Mass Customization.

To provide individualized yet affordable solutions, mass customization of home textiles is especially beneficial. Without having to deal with the high prices and

Table 1 Key differences of massive customization and mass customization [51]

Key difference	Massive customization	Mass customization
Degree of personalization	High, with textile solutions tailored to exact individual preferences	Moderate, with textile solutions tailored within a set framework of options
Cost and efficiency	Higher costs and longer production times due to the bespoke nature of the products	Lower costs and quicker production times due to the use of standardized components and processes
Production process	Often involves manual or small-scale production methods	Utilizes mass production techniques with integrated customization options

lengthy lead times of fully tailored products, homeowners can benefit from the aesthetic and practical advantages of customized fabrics [51].

6 Benefits of Custom Textiles

Numerous advantages come with custom textiles, including aesthetically pleasing and functional designs that are suited to a person's preferences and demands. Homeowners can create a distinctive look that adds to the character of their place by choosing custom designs. Custom textiles also meet functional needs like meeting certain material preferences or fitting non-standard measurements. When compared to off-the-shelf alternatives, these solutions frequently offer better quality and longevity. Furthermore, the process of tailoring textiles assures that each piece supports the overall home design concept and promotes creative freedom. All things considered, tailored fabrics improve interior spaces' aesthetic appeal and usefulness.

7 Case Studies of Textile Design in Interior Spaces

According to David's cross-sectoral expertise with mass customization in home textiles, established retailers with longer product life cycles have difficulties when the market shifts towards fashion-oriented goods. A study that was partially funded by the UK Department of Trade and Industry investigates the potential of digital printing for textiles to facilitate mass customization. David noted that managing color, using less black ink, adhering to industry standards, and standardizing processing conditions were the main technical difficulties. These problems were solved by advancements in technology, software, dyes, and inks. He also emphasized the difficulties in marketing, such as finding appropriate items, connecting with customers, and facilitating virtual collaboration between businesses. Mass customization is a paradigm change since traditional producers lack retail experience and retailers are used to keeping stock items. Despite these obstacles, businesses have discovered ways to sell digitally printed goods online and through high street merchants, creating jobs and spin-off businesses. David's research underscores that while technical challenges are manageable, marketing obstacles persist. Existing consumer channels have yet to fully embrace mass customization, but new digital avenues present significant opportunities. Niche businesses with specialized technologies can succeed, but new skills in marketing, brand management, and consumer access are essential. This case study emphasizes the need for cross-sectoral collaboration and innovation to achieve mass customization in the household textiles market [52].

Khalid's work demonstrates creative advancements in the creation of decorative indoor textiles, powered by technological advances and worldwide environmental considerations [53]. Focusing on functional applications, the research blends aesthetic appeal with health and vector protection, crucial in Pakistan's context of environmental conditions and vector-borne diseases. Using design thinking methodology, the study assessed user acceptance and manufacturing feasibility within Punjab's textile industry, ensuring desirability, feasibility, and viability.

A key innovation is custom home textiles functioning as mosquito repellents through vector protective finishes. These products offer aesthetic and functional benefits, catering to health and environmental needs. The study emphasizes the importance of accessibility and provides a needs analysis for integrating functional applications into home textiles. Conventional window textiles are easily accessible, while functional window textiles that possess properties like sound absorption, UV protection, or thermal insulation are less common and need to be produced by specialized companies. Traditional textiles prioritize aesthetic appeal, but functional textiles appeal to consumers seeking practical benefits such as energy efficiency and moisture resistance.

The study also addresses the higher manufacturing costs of custom functional textiles, involving additional stages and specialized materials. Custom functional textiles utilize technical fibers and advanced treatments, incorporating innovative approaches like nanotechnology and smart textiles for specific properties. Khalid concludes that designing custom functional and aesthetic textile products hinges on accessibility, presenting new opportunities and challenges. The revolutionary potential of incorporating modern technologies and environmentally friendly methods into interior textile design is highlighted by this case study [53].

Meiyu Yao's case study examines how design thinking can be applied to innovate elderly care products, with a particular focus on custom textiles. Design thinking, a human-centric approach, starts with empathy to identify complex problems and generate ideas. Chitosan, a natural antibacterial material, was proposed for developing custom skin-protective home textiles, addressing the needs and preferences of the elderly in care centers. The iterative process tackled spinning issues, studied biological and physical characteristics, and refined designs, sizes, patterns, and printing to meet user needs. After personalized beds, aprons, and handkerchiefs were successfully mass-produced, a 42-day pre-post-trial was carried out to assess utility and satisfaction. Feedback was generally positive, showing high satisfaction and potential health benefits. Design thinking took the place of conventional problem-solving techniques, emphasizing incremental releases and end users to sustain high satisfaction levels. The study identified elderly needs through empathy, highlighting that appropriate care textiles can significantly ease caregiver burdens and are better for elderly skin [54].

Ten Bhömer et al. developed novel solutions by combining several fields, such as fashion design, textile technology, material science, biology, and medicine [55]. Custom chitosan-cotton mix textiles were developed due to chitosan's antimicrobial qualities, which made it appropriate for use in hospitals and senior living facilities. Continuous discussions and site visits ensured accurate product specifications and designs preferred by the elderly. Objective tests confirmed the fabrics' quality and compliance with international standards. The trial at TWGHs (Tung Wah Group of Hospitals) elderly centers involved producing and testing 100 sets of custom bedding, aprons, and handkerchiefs, resulting in increased satisfaction and potential health improvements. The case study illustrates how design thinking may minimize challenges to product development and maximize advantages when it comes to the development of elderly textiles through in-depth user research and direct connection with end users [55].

8 Innovative Companies Which Offer a Range of Custom Textile Solutions

8.1 Camira

Based in West Yorkshire, Camira offers customized textiles for residential and commercial interiors, focusing on sustainable and high-performance fabrics to meet diverse aesthetic and functional needs. Founded in 1974, Camira provides a range of custom home textiles, including upholstery fabrics, curtains, and decorative panels. They emphasize sustainability and high performance, with extensive customization options to cater to various residential settings [56].

8.2 Eurojersey

Founded in 1960, EUROJERSEY specializes in high-quality Sensitive® fabrics, offering extensive customization and personalization services. They collaborated with Cifra to create innovative home textile solutions such as customizable upholstery, bed linens, and decorative items. Their focus is on combining advanced seamless technology with unique fabric designs to deliver high-performance and aesthetically pleasing home textiles [57].

8.3 Fuse

Founded in 2015, FUSE provides sustainable composite materials with a strong emphasis on customization. They specialize in eco-conscious home textiles, including upholstery fabrics and innovative textile solutions. FUSE's commitment to environmentally friendly production processes ensures that their customized products meet the highest standards of sustainability and performance [58].

8.4 TradeUNO

TradeUNO, co-founded by Sanjeev Goel and Ankur Aggarwal, specializes in providing a wide range of customized home textiles. A wide range of textiles, including cotton, silk, satin, and more than 150 other categories, are available from TradeUNO, guaranteeing quality, longevity, and a huge selection of designs and colors. Over 600 brands globally rely on their network for fabric supply, which serves both B2B and B2C sectors. With the help of TradeUNO's customization services, clients may choose from a variety of materials, like georgette, chanderi, jacquard, and more, made to their own tastes. With an emphasis on organic and sustainable solutions, including recycled fabrics, they provide household textiles like bedsheets, embroidery fabrics, and digital print materials. Customers can customize their orders to fit their individual design preferences using their user-friendly e-commerce platform, which makes use of cutting-edge digital tools and immersive 3D vision to create a custom experience. TradeUNO has established itself as a leader in the home textiles sector due to its dedication to sustainability and personalization, which enables them to produce goods that satisfy individual tastes while upholding high performance standards and environmentally responsible procedures [59].

9 Conclusion and Recommendation

Interior spaces benefit greatly from the use of textiles in both their aesthetic and practical features. Their versatility, which is seen in the variety of materials, designs, and applications they come in, makes them an effective tool for remodeling interior spaces in both homes and businesses. This chapter emphasizes the ways in which creative recycling of waste materials, the use of eco-friendly and sustainable textiles, and biophilic design concepts may all help create more sustainable and healthy living spaces. Technological developments in textile production, such as smart textile solutions and mass customization, present chances to satisfy contemporary design demands while tackling environmental issues. As essential components of interior design, textiles continue to develop by fusing innovation and tradition, encouraging coziness, sustainability, and individuality.

Designers, producers, and consumers must take innovative stances to fully utilize textiles in interior spaces. To start with, the usage of sustainable and eco-friendly textiles like bamboo, jute, linen, and organic wool should be given priority to lessen the influence on the environment. Furthermore, landfill contributions can be greatly reduced by using innovative techniques for recycling and upcycling textile waste into cutting-edge goods like furniture, rugs, and tiles. Second, to satisfy practical requirements in modern interior spaces, technical textile advancements—which offer qualities like UV protection, thermal insulation, and antibacterial finishes—should be included.

It is essential to educate customers about the advantages of eco-friendly and sustainable textiles. The primary aim should be to inform consumers about the effects of textiles on the environment, human health, and indoor air quality. To improve mental and physical well-being, designers should also embrace biophilic design principles, which incorporate natural components into places. Additionally, adopting mass customization techniques enables homeowners to design customized textile solutions that strike a balance between affordability and uniqueness. Finally, encouraging cooperation between sectors such as interior design, technology, and textile manufacturing will spur innovation and guarantee that textiles continue to satisfy changing aesthetic and environmental demands. The textile and interior design sectors may help create future spaces that are more aesthetically pleasing, sustainable, and healthy by putting these suggestions into practice.

References

1. Sage, C. (2014). *Design for the environment: An exploratory study on the processes that guide the design of interior textile products*. Colorado State University, ProQuest Dissertations & Theses.
2. Huppatz, D. J. (2010). The cave: Writing design history. *Journal of Writing in Creative Practice, 3*(2), 135–148.
3. Hasell, M. J., & Benhamou, R. (1988). Interior design: A dynamic systems view. *Journal of Interior Design Education and Research, 14*(2), 13–22. https://doi.org/10.1111/j.1939-1668.1988.tb00124.x
4. Nielson, K. J. (2007). *Interior textiles: Fabrics, application, and historic style*. John Wiley & Sons.
5. Farooq, S., & Kamal, M. A. (2021). Exploration of fabric typology for interior furnishings with reference to regional textile products of Pakistan. *Civil Engineering and Architecture, 9*(1), 166–175.
6. Cheng, J. P. (2003). *Home fashion: A concept of creating home furnishing products using fashion theory and design process* (Master's thesis). College of Design, Art, Architecture, and Planning, University of Minnesota.
7. Canavan, K. (2015). Applications of textile products. In *Textiles and fashion* (pp. 531–545). Woodhead Publishing. https://doi.org/10.1016/b978-1-84569-931-4.00021-0
8. Qiao, L. G., Feng, S., Rui, B. H., & Xiao, Y. L. (2012). Innovative use of wood materials in modern home design. In *Applied Mechanics and Materials* (Vol. 204–208, pp. 4156–4160). https://doi.org/10.4028/www.scientific.net/AMM.204-208.4156

9. Lee, S., & Buck, L. (2020). Sustainable design approaches using waste furniture materials for design students. In *International Conference on Engineering and Product Design Education*. VIA University College.
10. Chen, Y., & Peng, H. (2017). Research on creative designing of home decoration by using waste wood. In *Proceedings of the 3rd International Conference on Social Science and Management (ICSSM 2017)* (pp. 678–683).
11. Calovi, M., & Rossi, S. (2023). From wood waste to wood protection: New application of black bio-renewable water-based dispersions as pigment for bio-based wood paint. *Progress in Organic Coatings, 180*, 107577. https://doi.org/10.1016/j.porgcoat.2023.107577
12. Missia, D. A., Demetriou, E., Michael, N., Tolis, E. I., & Bartzis, J. G. (2010). Indoor exposure from building materials: A field study. *Atmospheric Environment, 44*(38), 4388–4395.
13. Bartzis, J., Wolkoff, P., Stranger, M., Efthimiou, G., Tolis, E. I., Maes, F., Nørgaard, A. W., Ventura, G., Kalimeri, K. K., Goelen, E., et al. (2015). On organic emissions testing from indoor consumer products' use. *Journal of Hazardous Materials, 285*, 37–45.
14. Cacho, C., Ventura Silva, G., Martins, A. O., Fernandes, E. O., Saraga, D. E., Dimitroulopoulou, C., Bartzis, J. G., Rembges, D., Barrero-Moreno, J., & Kotzias, D. (2013). Air pollutants in office environments and emissions from electronic equipment: A review. *Fresenius Environmental Bulletin, 22*, 2488–2497.
15. Nandan, A., Siddiqui, N. A., & Kumar, P. (2020). Estimation of indoor air pollutant during photocopy/printing operation: A computational fluid dynamics (CFD)-based study. *Environmental Geochemistry and Health, 42*, 3543–3573.
16. Li, A. J., Pal, V. K., & Kannan, K. (2021). A review of environmental occurrence, toxicity, biotransformation, and biomonitoring of volatile organic compounds. *Environmental Chemistry and Ecotoxicology, 3*, 91–116.
17. Dallinga, J. W., Robroeks, C. M. H. H. T., Van Berkel, J. J. B. N., Moonen, E. J. C., Godschalk, R. W. L., Jöbsis, Q., & Van Schooten, F. J. (2010). Volatile organic compounds in exhaled breath as a diagnostic tool for asthma in children. *Clinical & Experimental Allergy, 40*(1), 68–76.
18. Seguel, J. M., Merrill, R., Seguel, D., & Campagna, A. C. (2017). Indoor air quality. *American Journal of Lifestyle Medicine, 11*(4), 284–295.
19. Donatello, S., Moons, H., & Wolf, O. (2017). *Revision of EU ecolabel criteria for furniture products*. European Commission Technical Report. http://ec.europa.eu/environment/ecolabel/documents/technical_report_furniture.pdf
20. Kansai Paint. (2009). *Water-based paint composition* (US Patent Application Number US 2009/0099298)
21. Kansai Paint. (1994). *Aqueous color paint compositions and coating method using same* (US Patent No. 5,322,865)
22. Nippon Paint. (1998). *Heat-curable, water-dispersible resin composition, production thereof, water-based paint composition, method of coating and coated article* (US Patent No. 5,747,558)
23. Nippon Paint. (1999). *Method of coating using pigment-containing water-based paint composition* (US Patent No. 5,972,425)
24. Miao, J., Wong, K., Wei, L., Hang, N. S., Hang Keung, L., Ho, S. K., Ho, I. Y. M., Luk, R. K. C., Cai, L., Cheng, C., Tsang, G. Y. Y., & Lee, P. W. (2012). Nanomaterials applications in "green" functional coatings. *Nanotechnology Perceptions, 8*, 181–189.
25. Wilson, E. O. (1984). *Biophilia*. Harvard University Press.
26. Kellert, S. R. (2005). *Building for life: Designing and understanding the human-nature connection*. Island Press.
27. Kellert, S. R. (2008). Dimensions, elements, and attributes of biophilic design. In J. Heerwagen & M. Mador (Eds.), *Biophilic design: The theory, science, and practice of bringing buildings to life* (pp. 3–19). Wiley.
28. James, S. L., Abate, D., Abate, K. H., Abay, S. M., Abbafati, C., Abbasi, N., Abbastabar, H., Abd-Allah, F., Abdela, J., Abdelalim, A., et al. (2018). Global, regional, and national incidence, prevalence, and years lived with disability for 354 diseases and injuries for 195 coun-

tries and territories, 1990–2017: A systematic analysis for the Global Burden of Disease Study 2017. *The Lancet, 392*, 1789–1858.
29. Feigin, V. L., Vos, T., Nichols, E., Owolabi, O. M., Carroll, W. M., Dichgans, M., Deuschl, G., Parmar, P., Brainin, M., & Murray, C. (2019). The global burden of neurological disorders: Translating evidence into policy. *Lancet Neurology, 19*(3), 255–265.
30. Booth, F. W., Roberts, C. K., Thyfault, J. P., Ruegsegger, G. N., & Toedebusch, R. G. (2017). Role of inactivity in chronic diseases: Evolutionary insight and pathophysiological mechanisms. *Physiological Reviews, 97*(4), 1351–1402.
31. Brasche, S., & Bischof, W. (2005). Daily time spent indoors in German homes: Baseline data for the assessment of indoor exposure of German occupants. *International Journal of Hygiene and Environmental Health, 208*(4), 247–253.
32. Pears, M., Kola-Palmer, S., & De Azevedo, L. B. (2021). The impact of sitting time and physical activity on mental health during COVID-19 lockdown. *Sport Sciences for Health, 17*(4), 1–13.
33. Manger, S. (2019). Lifestyle interventions for mental health. *Australian Journal of General Practice, 48*(11), 670–673.
34. Dean, E., & Söderlund, A. (2015). What is the role of lifestyle behaviour change associated with non-communicable disease risk in managing musculoskeletal health conditions with special reference to chronic pain? *BMC Musculoskeletal Disorders, 16*(1), 1–7.
35. Bulaj, G., Clark, J., Ebrahimi, M., & Bald, E. (2021). From precision metapharmacology to patient empowerment: Delivery of self-care practices for epilepsy, pain, depression, and cancer using digital health technologies. *Frontiers in Pharmacology, 12*, 612602.
36. Huntsman, D. D., & Bulaj, G. (2022). Healthy dwelling: Design of biophilic interior environments fostering self-care practices for people living with migraines, chronic pain, and depression. *International Journal of Environmental Research and Public Health, 19*(4), 2248. https://doi.org/10.3390/ijerph19042248
37. Zhong, W., Schröder, T., & Bekkering, J. (2022). Biophilic design in architecture and its contributions to health, well-being, and sustainability: A critical review. *Frontiers of Architectural Research, 11*(1), 114–141.
38. Shakour, S., & Hammash, A. (2023). The environmental and health impact of eco-friendly textiles in the interior space. *SVU-International Journal of Engineering Sciences and Applications, 4*(1), 35–40.
39. Keune, S. (2018). *On textile farming: Seeds as material for textile design* (Master's thesis). The Swedish School of Textiles, University of Boras. https://hb.diva-portal.org/smash/get/diva2:1364974/FULLTEXT02.pdf
40. Kaur, L. (2021). Batik pattern and natural dyes in home décor using contemporary art. *Review Journal of Philosophy and Social Science, 46*(1), 151–160. https://doi.org/10.31995/rjpss.2021.v46i01.018
41. Babel, S., & Sachihar, L. (2020). Designing cushions picking inspiration from traditional folk painting: Sanjhi. *International Journal of Science and Research, 9*(1), 1106–1108.
42. Gupta, M., & Gangwar, S. (2016). Adaptation of designs for textile products inspired by Madhubani painting. *International Journal of Research—Granthaalayah, 4*(5), 115–125.
43. France, L. (2022). *Development of home décor from post-consumer waste by implementing upcycling and transformable methods* (Master's thesis). St. Teresa's College (Autonomous), Department of Fashion Designing, Ernakulam. http://117.239.78.102:8080/jspui/bitstream/123456789/2249/1/LISIEUX%20FRANCE.pdf
44. Sisodia, N., & Parmar, M. S. (2022). Converting textile waste into designer wall and floor tiles: A new approach to recycle textile waste. In *Sustainable approaches in textiles and fashion: Circular economy and microplastic pollution* (pp. 149–164). Springer Singapore.
45. Jamshaid, H., Shah, A., Shoaib, M., & Mishra, R. J. (2024). Recycled-textile-waste-based sustainable bricks: A mechanical, thermal, and qualitative life cycle overview. *Sustainability, 16*(10), 4036. https://doi.org/10.3390/su16104036

46. Nielson, K. J. (2007). Sustainable interior textiles. In *Interior textiles: Fabrics, application, and historic style* (pp. 23–34). John Wiley & Sons.
47. Spallek, J., & Krause, D. (2016). Process types of customization and personalization in design for additive manufacturing applied to vascular models. *Procedia CIRP, 50*, 281–286.
48. Borstrock, S. (2018). Personalization, customization and bespoke: Increasing the product offer. *Journal of Design, Business and Society, 4*(2), 171–187.
49. Srivastava, A. (2024) *AI in luxury fashion: Redefining creativity and customer experience*. https://www.fibre2fashion.com/industry-article/10054/ai-in-luxury-fashion-redefining-creativity-and-customer-experience
50. Kusnezov, X. (2012). *Custom-made fashion: Customized fashion collection*
51. Kolarevic, B., & Duarte, J. P. (Eds.). (2018). *Mass customization and design democratization*. Routledge.
52. Tyler, & David, J. (2006). Cross-sectoral experiences with mass customization. In *Fashion in the Digital Age: 8th Annual IFFTI Conference*. North Carolina State University.
53. Khalid, A., Malik, M. H., Farooq, S., & Khan, S. (2024). Designing functional and aesthetic home textiles for mosquito control in the indoor built environment: A review of needs assessment and opportunities in Punjab. *Journal of Arts and Social Sciences, 11*(1), 65–77.
54. Yao, M., & Li, L. (2022). Design thinking applied to home textiles innovation: A case study in an elderly centre in Hong Kong. *Designs, 6*(3), 49. https://doi.org/10.3390/designs6030049
55. Ten Bhömer, M., Tomico, O., & Wensveen, S. (2016). Designing ultra-personalized embodied smart textile services for well-being. In *Advances in smart medical textiles* (pp. 155–175). Woodhead Publishing.
56. Camira Fabrics. (n.d.). *Sustainable and customizable textile solutions*. https://www.camirafabrics.com/en
57. Sensitive Fabrics. (n.d.). *Innovative fabric solutions*. https://www.sensitivefabrics.it
58. Fuse Composites. (n.d.). *Advanced sustainable composite materials*. https://en.fuse-composite.com
59. TradeUNO. (n.d.). *Custom textiles and fabrics*. https://www.tradeuno.com

A Vision of Textile Design in the Field of Industry Professionals

J. Jaisri and S. Balaji

1 Introduction

Textile making is a job that never stops changing. It mixes art, high-tech ways, and attention to detail to make fabrics. But these fabrics aren't only nice to look at; they can also change how we think about clothing, decorating our homes, and the culture around us. Designing fabrics is a job that asks for imagination and exactness with a sense of balance between the look, use, and how it affects our planet. This article points out that people in this field make materials come alive. For this, they need a good eye for design and a deep knowledge of materials, fibres, and procedures like weaving, knitting, and printing with machines [1]. They get their ideas from many places, like nature, history, and modern-day art, turning these ideas into fabric patterns and textures that show how they see the world [2].

Textile design is all about trying new things. To stay in the game, designers must be open to using newer methods and materials. New tech textile materials with digital printing and earth-friendly fibres have opened up a world of possibilities. Expert designers have told us that being good at using this tech is just as important as having a creative mind. It helps turn unique ideas into beautiful and useful products. These designers stress the need to learn and adapt constantly to handle the latest tools and tricks. This way, they stay important in a field that's becoming more and more shaped by tech upgrades [3].

J. Jaisri (✉)
Department of Costume Design and Fashion, School of Creative Sciences, Nehru Arts and Science College, Thirumalayampalayam, Coimbatore, Tamil Nadu, India

S. Balaji
Department of Commerce, School of Commerce, Nehru Arts and Science College, Thirumalayampalayam, Coimbatore, Tamil Nadu, India

Sadhna et al. (eds.), *The Art and Craft of Modern Textile Design*, Sustainable Textiles: Production, Processing, Manufacturing & Chemistry,
https://doi.org/10.1007/978-3-031-86797-2_10

Textile designers who invent new things face tough situations. They have to be strong and smart. Customers want good quality, useful, and earth-friendly products. This puts a lot of pressure on the designers. They need to use earth-friendly ways without losing their creativity, and it also shows that designers often work with environmental experts and scientists. They do this to create earth-friendly textiles. They aim to meet the needs of the industry and customers [4]. Working with others not only helps the design process but also shows the power of textile design. It can help the environment and change the story people believe about it. They have worked with big fashion brands and also worked on projects to lessen harm to the environment. These show that textile design can bring about real change.

Fashion industry experts offer great advice for those starting in textile design [5]. These tips help the new innovators to build strong skills. Senior designers push fresh talents to learn many skills in both imaginative and technical areas [6]. They believe beginners should get a strong education, create a good portfolio, and keep up with the latest styles to do well [7]. They also say that to have a lasting career, designers need to be unique and have a great vision [8]. They think the best careers are founded on a promise to bring something new, be it through new studies or unique-looking and impactful designs [9]. This study digs deep into textile design, showing the hard work, knowledge, and passion that fuel this area. From these professionals' life stories, readers learn about the blend of creativity and technical skill needed in textile design. They also see its big role in shaping fashion, culture, and eco-friendly ways. Textile design is more than just making pretty things. It's about creating fabrics that tell stories, stir feelings, and lead to good changes, showing the big effect this field can have on the world.

2 Sustainable Innovations in Textile Design

Textile design has evolved from a traditional craft to a field where art meets Technology over the last few years. The tension between aesthetics and technology is, for Silicon Valley, literally producing a new kind of fabric with the potential to make an impact on fashion but also on bigger issues [10]. When new tools and techniques have become available, designers have sought to stretch the limits of patterning, fabric texture, and sustainability to encapsulate what modern textile design can be. Textile design, as an ever-evolving frontier of the place where man meets machine, strikes an unparalleled balance between both fields, enabling human creativity to flourish, while allowing tactile sensations and tactile development to create beautiful fabrics characterized by superior functionality. Figure 1 shows the key factors influencing a textile designer's creative process.

Textile designers' methods have shifted with technological advances, from digital printing to automated weaving or fabric made with 3D printers [11]. CAD software enables designers to play around with different patterns, textures and colours instantly, and, in turn, visualizations of complex ideas evolved quickly even before

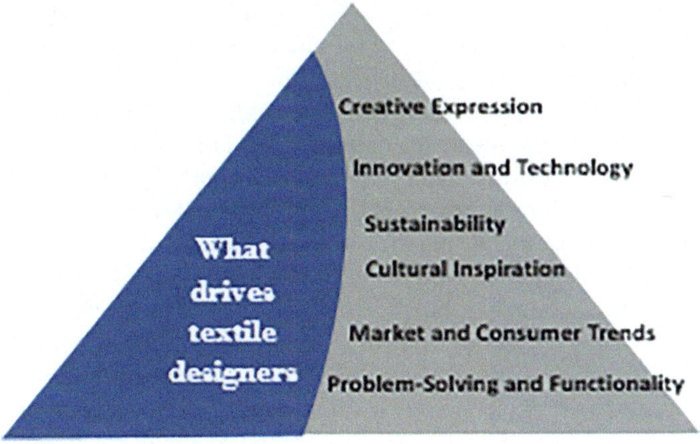

Fig. 1 The various drives of textile designers [11]

the production. CAD has sped up the design process while also decreasing errors by allowing designers to perfect every aspect of the product digitally, from colour combinations to fabric layouts. Intricate designs can now be laser cut from materials and digitally printed onto fabrics with a precision never achieved hands-on. For example, 3D knitting and weaving technologies enable designers to add textures and dimensions to fabrics, moving beyond flat patterns. It offers the ability to play around with different types of forms and designs to generate textiles that serve functions and motions, for example, stretch, insulation, or breathability, without sacrificing the visual appearance of the fabric [12].

Among the most important intersections of creativity and technology is sustainability [13]. Due to its resource-intensive manufacturing processes and the high demand for raw materials and energy, textile design is one of the highest contributors to pollution on a global scale. Now sustainable materials and energy-efficient techniques are finding their way into textile design innovations to address these challenges. Digital printing methods, for instance, decrease the amount of water used compared to traditional dyeing methods, while old materials such as plastic bottles are being cobbled into fresh fabrics made to be new.

Biotechnology also started playing a role in textile design, with designers experimenting with lab-grown fibres that behave like traditional fabrics but with a significantly lower environmental cost [14]. This latest wave of sustainable innovation includes mushroom leather, algae-based textiles, and biodegradable fabrics [15]. Textile professionals are looking at how science meets design to utilize the potential of technology in aiming at quality bio fabrics without compensating for the quality or look and feel of the fabrics [16].

Textile design is becoming infused with augmented reality (AR) and virtual reality (VR) augmenters [7]. This enables designers to simulate showrooms and fabrics virtually, enabling their clients to preview the visualization and feeling of the textiles even before it is made [15]. Through AR and VR, designers are able to communicate their ideas and visions in an interactive way, allowing for a better experience and a deeper insight from the client side of the product potential [5]. These online applications particularly help in controlling wastage [17]. This allows designers to make several proposals without the need for a physical sample, avoiding waste of materials and resources in the process [18]. Customisation is aided by virtual fitting rooms, allowing clients to "try on" designs in fashion and home décor [19]. Such a digital-driven approach is leading to visualising textile possibilities in a more efficient and sustainable manner. While this intersection of textile design and modern technology is changing the way we create fabrics, its roots are still firmly planted in traditional craftsmanship. So many designers combine heritage tools with modern-day weaving elements from time-tested designs with methods that have been around for centuries into contemporary work. The fabrics carry their own characters, often in ways that technology cannot replicate, derived from hand-loomed, traditional dyeing methods and regional weaving techniques. Through a careful awareness of cultural heritage and its integration within a digital context, designers generate textiles that build upon tradition while adapting to future-oriented methodologies [20, 21].

With developments in artificial intelligence (AI) and machine learning slowly penetrating the industry, the future of textile design has never looked brighter. Using machine learning to monitor trends and predict consumer preferences, AI algorithms can guide designers as they make creative decisions. The AI-enabled design also simplifies designers to create new patterns. Moreover, the field of textile design is gaining a new frontier with wearable technology. For instance, fabrics containing sensors that measure body temperature, physical movement, or the degree of hydration of the body are already incorporated into sportswear, medical garments and garments for fashion purposes. This blend of fashion and function has come to be shorthand for cutting-edge developments in textile design.

3 Textile Design for Sustainability

While every field is built on a foundation of theoretical knowledge, practical exposure and application in the field become equally significant. This is where the expertise of industry professionals can help by sharing best practices, effective approaches, and mistakes to avoid. Such conversations have an enormous amount of expertise behind them, and we can learn much about common trends and common challenges from guilds of professionals from a number of successful categories and make sense of success.

- What are the most exciting trends in textile design and what are some innovations that are changing the industry?
- How is the industry adapting toward sustainability? What difficulties do companies encounter in implementing them?

- What effect does the emergence of new technologies such as AI and digital printing have on textile design and production? Are there impediments to the adoption of these?
- Effect of consumer demand on impact textile design.
- How do social media and fast fashion shows shape designs to be reflected?
- What does the future hold for textile design, according to experts? What trends or technologies do they see coming?

Lululemon and Under Armour have been testing textiles that respond to body temperature, sweat, and other physiological changes. Even medical and functional wearables will follow this trend, combining functionality and fashion [22].

Some designers that reflect this are Gabriela Hearst and Mara Hoffman, who use dye techniques from culturally sustainable and natural sources such as plants and minerals that produce less toxic waste than the dyes that are synthesized to make clothes [23].

Brands like Levi's, Everlane, and Outerknown target use of organic cotton, hemp and recycled polyester. Yet cannot manage to get on hands of these materials sustainably both and in quantity due to the price tag and the supply chain being more constrained [24].

Textiles titans such as Adidas and Nike are now turning to waterless dyeing and closed-loop recycling to minimise waste. But scaling these innovations internationally is challenging—it requires massive resources to build infrastructure and train operators [25].

Designers like Christopher Kane have incorporated digital printing, which allows for greater detail with less fabric, and AI is now offering the potential for design ideas and insights into trends [26]. Zara and H&M have even begun employing AI to shorten the design-to-production process by allowing AI to dictate what consumers will purchase, and consumers are choosing brands that provide information about sourcing, production, and labour [23]. Companies such as Everlane and Reformation are leading the way with communication around cost, sourcing, and sustainability [27].

Brands, including the likes of Nike and Burberry, have adopted personalized features to appeal to a more individualistic and expressive consumer base. The trend suggests a transition from mass items to bespoke, customized designs.

Online platforms such as Instagram and TikTok move trends along; we perpetually crave a fresh take on design. Adam: Social media allows Gucci and Balenciaga to use the influencer marketing piece to sell new collections and trends, but they also have to provide what the people want at a very fast pace, which loses sustainability in the process.

Brands like SHEIN and Zara make high-fashion designs quickly and cheaply. This makes fashion touch everyone, but also causes a lot of waste because it's made so fast. We might start to make clothes from things that grow in a lab, like Mylo, a type of leather made from fungi. Big brands like Adidas and Stella McCartney are involved in this effort. Using these new materials could change how we make clothes and not have to use animal products or fake materials. Polygiene Technologies believes these changes are good. They can make clothes look nicer, feel more comfortable, and work better.

1. **Antimicrobial additives:** They keep clothes fresh longer between washes, offering a new angle for designs.
2. **Biodegradable textiles:** These materials break down naturally, making clothes more planet-kind.
3. **3Upcycling:** Companies like Patagonia, through their Worn Wear initiative, offer credits for trading in old gear that they remake and resell.
4. **3D printing:** It makes complex designs and awesomely functional features a reality through cool manufacturing techniques.
5. **Smart textiles:** They come with sensors to keep tabs on health stats or changes in the environment and provide immediate feedback to the users.
6. **Anti-odour clothing:** Products like Polygiene StayFresh keep sweat and bacteria-induced smells at bay, while OdorCrunch™ tackles smells from the surroundings.
7. **Nanotechnology:** This has the potential to create self-cleaning fabrics, lowering how often we need to wash them and lengthening their life.
8. Chemical recycling textiles can get broken down into basic materials through closed-loop systems like chemical recycling. This reduces dependency on unused resources.
9. Hydrophobic coatings repel water and dirt, so you don't need to wash as much. Virtual reality (VR) and augmented reality (AR) make customer experiences better with virtual try-ons.
10. Artificial intelligence (AI) makes design and manufacturing processes more efficient.
11. Synthetic biology helps in creating materials in labs for eco-friendly production.
12. Advanced robotics improves the efficiency and precision of manufacturing.
13. Blockchain technology makes supply chains more open. This ensures that all production happens ethically.

Polygiene is on a mission. They're digging into these changes and guiding brands on how to meet the changing needs of their buyers. At the same time, they promote earth-friendly habits in the cloth-making trade. Now, let's talk about the experts in this field. They each have a ton of firsthand experience from their line of work. Their opinions are built on victories and losses and they bring us the hidden side of every tale. This view is vital for ticking off every box on our to-do list. Having discussions with seasoned professionals and researchers gives us a better look at the nooks and crannies of our industry. This knowledge goes beyond basic learning routes. First, a chat with industry pros can help you be the first to know about emerging trends. These experts can feel the direction the market is headed. This sort of foresightedness can be super crucial for businesses and people who are keen to increase their profits [28].

4 Key Takeaways from Expert Conversations

Flexibility and Lifelong Learning: The professionals in this field emphasize flexibility, knowledge-building, and adaptability. When it comes to the economic world, stagnation is not an option for most industries owing to the rapid pace of change. Experts say it's important to keep learning which could mean attending formal seminars, informal workshops on open-source tools or even just reading up on your own. Such flexibility not only allows new skills to be learned but also what they need to do and how they need to think [29].

4.1 Getting Ahead

Learning to climb the ladder of success is not a one-and-done scenario. Based on recent conversations with some long-time industry veterans, resilience and persistence might be the key difference makers. It is common to read from experts the stories of its early failures and setbacks, how they all had to overcome difficulties and learn from them in order to achieve success especially with the current unpredictable global environment. Especially with the unpredictability of the global environment today, this [28].

4.2 Guiding, Edifying the Future

Getting guidance from experts will help guide the next generation of professionals. Such advice can help give younger adults a sense of confidence and guidance into what is a new and daunting experience. Mentorship programs (whether formal or informal) can also be immensely beneficial [30].

4.3 Policy and Standards

Experts greatly inflect the policy and standards of their field, ensuring that growth and innovation continue through decennial evolution. Their involvement can have a great impact on regulatory policy settings, industry standards, and best practices for the sake of an agile enterprise culture. The involvement of these professionals offers a broader perspective on the environment of industries [28].

4.4 Networking and Collaboration

Networking and collaboration persist as one of the remaining essential ones. A strong network tends to provide you with opportunities that you otherwise would not have had a chance to experience. Collaboration across industries and functions can create novelty [30] and new business.

To foster professional development and move the industry forward, industry experts need to have real-world conversations. Having to pull lines of communication between these diverse groups enables the practitioners in the field to see the whole picture and find paths much more easily. As industries continue to change and rethink old patterns of work, the relevance of subject matter experts will only continue to grow further, justifying the need to remember talking to the people that are at the forefront of the field.

5 The Creative Process of Industry Experts and Fashion Designers

The creative process is infinitely complex and fascinating, especially in the fashion world. Professionals navigate the fine line between artistic expression and commerce, transforming abstract thoughts into concrete objects that set trends and influence consumer behaviour. Below, we take a deeper look into the behind-the-scenes details of how these pros create magic within their design processes by sharing insights on methods, inspirations, and roadblocks encountered on their journey in Fig. 2.

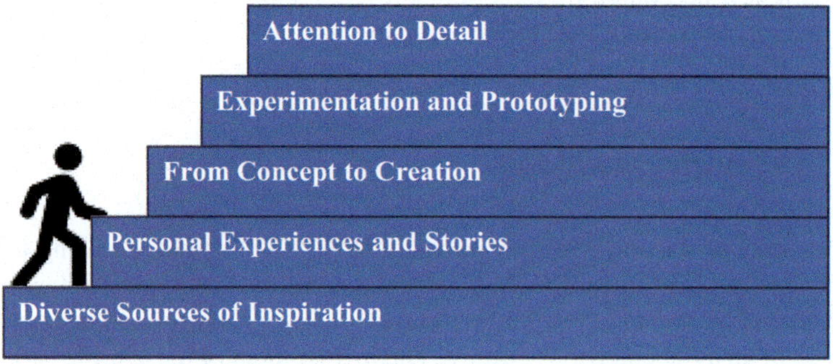

Fig. 2 Shows the creative process of industry experts [1]

5.1 *Diverse Sources of Inspiration*

Inspiration for fashion designers can stem from various sources. Nature, art, history, and everyday life often fuel creativity [1]. Renowned designers like Alexander McQueen delved deep into history and heritage, reviving the past through groundbreaking collections. Industry experts advocate for being conscious and alert to the surrounding world, as creativity can strike at any moment—even in mundane spaces.

5.2 *Personal Experiences and Stories*

Each designer possesses a unique identity, and this should be reflected in their design process. Many jewellery designers derive inspiration from their life stories, creating deeply personalized objects. This emotional resonance appeals to contemporary consumers seeking a connection with the brands and products they choose. Designers infuse their work with personal narratives, resulting in offerings that tell stories, not just showcase clothing.

5.3 *Concept Development*

A strong vision guides the transition from idea to physical reality. Designers often begin with sketches or mood boards that collect fabric swatches, colour, and theme-based ideas. The conceptual stage is crucial as it allows designers to explore ideas and refine their vision. Collaboration is also vital during this phase; input from team members like pattern makers and textile experts is invaluable in shaping the final design.

5.4 *Experimentation and Prototyping*

Experimentation is essential in the creative process of fashion design [9]. This phase allows designers to test different materials, techniques, and silhouettes that align with their story. Creating prototypes or sample garments enables designers to evaluate the viability and appearance of their ideas. This iterative approach often involves hours of revision, as designers meticulously refine each detail [31].

5.5 Attention to Detail

Successful fashion designers pay careful attention to even the tiniest of details. Every aspect, including the fabric used to construct garments, should be consistent from ideation through to stitching and the finished product. As many in the industry often remind us, the devil is in the details; even small elements that may be imperceptible during a handshake can become glaring during mass production. Attention to detail ensures that each piece not only looks good visually but also functions well on the body of whoever wears it [3, 4].

5.6 Balancing Creativity and Commercial Viability

For fashion designers, one of the biggest challenges is striking a balance between creativity and commercial viability. How creative can designers get before their work becomes too commercial? While innovation and artistic expression are paramount, designers must also stay attuned to market trends and consumer preferences. This balance ensures that collections are both unique and saleable. Scholars and practitioners emphasize the critical need for fashion firms to understand their target market while finding ways to shape creativity around established notions of consumer demand [1, 31]. The fashion industry faces a growing sustainability challenge. Designers are tasked with creating beautiful and innovative products while minimizing their environmental impact [32]. This means selecting sustainable materials, employing eco-friendly production methods, and considering the entire lifecycle of the garments. An increasing number of designers are integrating sustainability into their practices and advocating for reform within the fashion industry [32].

6 Challenges in Textile Design

6.1 Sustainability

One of the biggest challenges facing the textile design industry today is sustainability. The textile industry is known for its significant environmental impact due to water usage, the application of chemicals (such as dyes and finishes), and waste generation. Designers must discover eco-friendly materials and production methods for nearly every aspect of their work. In terms of sustainability, they aim to adopt better practices, such as utilizing organic farming for raw materials and minimizing water usage, all while aligning with zero-waste systems through their suppliers [32, 33].

6.2 *Technological Integration*

The convergence of technology and textile design presents numerous opportunities, but it also introduces several challenges. For many designers, the need to navigate complex software and machinery to bring their ideas to life is a significant hurdle. For instance, digital printing enables the creation of intricate, customized designs tailored to specific projects; however, it requires proficiency in digital tools and techniques [29, 34]. Additionally, the rapid pace of proprietary innovation necessitates continuous learning and adaptation to keep up with new technologies [7, 28].

6.3 *Making a Strong Portfolio*

Your future employers or clients always judge you by the work you show them. So, your portfolio must be well-made to show all your potential [35]. It should showcase your range, encompassing all kinds of work and all aspects of your personality and skills. Include sketching work, digital designs, images or fabric swatches, and photos of finished products. Your portfolio should be attractive and organized, and, of course, it should always be up to date [36].

Highlight your unique strengths. If you excel at photography, consider including your photos in the portfolio. Whether it's a special technique, innovation in the use of materials, or a distinctive style, your portfolio should tell your story. The progression should be smooth, with photos arranged sequentially from your earlier, less polished works to your more refined, stylistically cohesive pieces [36].

6.4 *Networking and Professional Development*

Networking is crucial for a career in the textile industry. Engaging in this process requires attending industry events, joining relevant organizations, and participating in forums or social network groups. Numerous opportunities exist for building relationships with other designers, professionals, and potential clients, some of which may lead to interesting job offers, joint projects, or mentorship [7, 30, 32].

Mentorship is extremely beneficial for aspiring textile designers. Finding an experienced practitioner who can offer advice and support is vital. This assistance will be essential for your career development and will help you make informed decisions on various matters. Many associations, schools, and universities offer formal mentorship programs.

The textile industry is dynamic and continuously evolving with new technologies, techniques, and design approaches. To stay informed, utilize various tools, such as reading industry reports, following influential designers and companies, subscribing to industry newsletters, and expanding your network of contacts. All these efforts will help you understand current trends and customer demands, which is crucial for designing innovative and marketable.

As global awareness of environmental issues and sustainable development principles grows, the textile industry is gradually adapting to this trend. As a future designer, a significant step toward understanding the industry specifics involves exploring sustainable materials and eco-friendly technologies [35].

6.5 Dealing with Rejection

In any creative career, rejection is inevitable [37]. However, it should be viewed as an opportunity for improvement. When facing rejection, use it as motivation for further work. Seek constructive feedback, analyse your strengths and weaknesses, refine your skills, and try again. Persistence and a positive outlook are essential for success in the textile business [38].

6.6 Coping with Creativity and Commerce Issues

While it's crucial to follow your creative vision, maintaining a career in this field requires balancing creativity with commercial demands. To build a successful career, develop your marketing knowledge to produce goods that appeal to a broad audience [33]. If you aspire to a particular style, be mindful of your target customer group's demands.

7 Conclusion

Textile design is a growing field; it's like a tree that blends art and technology and this mix shapes everything from what we wear to how we decorate our rooms. It adds beauty and usefulness to our everyday world.

Think of it as a meeting point between imagination and technical skill. Designers get their ideas from things like nature, cultural heritage, and modern art [1]. Technology has changed this field a lot over the years. New tools and clever ways of making things have greatly changed how fabrics get designed, made, and delivered to people [39]. However, unfamiliar things also bring new challenges. Designers today must consider the environment, be quick to add new technology, and continuously meet changing customer needs. Nowadays, making things green isn't just trendy. It's a need, especially for people who design materials. People care about Earth more now [12]. So, designers need to choose Earth-friendly methods. Think of a designer's job as telling the story of life. It starts with the basic stuff and ends with throwing away. What's their goal? Keep Earth clean [32]. They make earth-friendly fibres, use less water and power, and remake or boost old stuff [40]. Designers are partnering with green experts and material wizards. They want to make materials that are not only cool to see but also meet hard-to-get green norms [4]. Tech and material design can be a gift and a brain-teaser at the same time. Tools like 3D designs and computerized printing help designers make fancy patterns right, and they also trim garbage [40]. This has allowed even small squads or single designers to use the modern tech that only large makers used to have. But, with tech going so fast, designers need to keep growing and adding to what they know. Always learning is crucial for those in this job [3].

Designing textiles isn't just about creativity or knowing the technical stuff. You also have to understand what's happening in the market and what people want. To keep up in this tough field, designers need a lot of skills. They need to be artistic, sure, but they also need to know how to make sense of market data [7]. As what people like changes, designers have to change too. But it's important to keep their own unique style in the mix. This is really important if they want to make a big impact in this field. That might mean creating patterns nobody's seen before, using earth-friendly materials, or finding new ways to use the latest tech [41]. There's still so much potential in textile design. If it embraces being green, keeps up with tech advances, and works with other fields, it could make fabrics that are beautiful, practical and good for the planet. These can make our surroundings and lives a lot more meaningful. This piece shines a light on all the steps, challenges, and chances in textile design. It's a field that demands both technical know-how and creativity. Looking at it this way, we can see how much textile design plays a part in our culture, way of life, and environment.

References

1. Hargadon, A., & Sutton, R. I. (1997). Technology and the culture of innovation: Lessons from the textile industry. *Journal of Business Research, 52*(4), 387–395.
2. Hargadon, A., & Sutton, R. I. (1997). Technology brokering and innovation in a product development firm. *Administrative Science Quarterly, 42*(4), 716–749. https://doi.org/10.2307/2393655
3. Farrer, M., & Harris, L. (2017). The role of attention to detail in textile and fashion design. *Fashion Practice, 9*(1), 37–51.
4. Roberts, C., & Zhang, H. (2024). Biodegradable textiles: Merging tradition with innovation. *Sustainable Textiles Journal, 19*(1), 33–50.
5. Taylor, R. (2019). *Textile design essentials: A guide for students and professionals*. Laurence King Publishing.
6. Williams, J. (2012). *The art and science of textile design*. Rizzoli International Publications.
7. Cope, R., & Jack, S. (2010). The fabric of creativity: Innovative textile design. *Textile Journal, 15*(3), 56–68.

8. Cope, J., & Jack, S. (2010). *Managing innovation in the textile industry*. Routledge.
9. Smith, J. (2019). The importance of lifelong learning in the modern workforce. *Journal of Professional Development, 31*(1), 34–42. https://doi.org/10.1080/09528222.2019.1622187
10. Hsu, C.-H., & Chang, Y.-T. (2017). The evolving role of textile design in sustainability. *International Journal of Textile and Fashion, 10*(2), 123–136.
11. Nour, M. (2014). Mentorship and its impact on career growth: Insights for young professionals. *Career Development Quarterly, 62*(2), 163–175. https://doi.org/10.1002/j.2161-0045.2014.00074.x
12. Russell, M. (2018). The intersection of technology and textile design. *Journal of Modern Design, 15*(1), 33–42.
13. Allen, R., & Young, T. (2023). *Sustainable textiles: Innovations and practices*. Textile Futures Press.
14. Shaffer, L., & Purdy Brown, T. (2012). Collaborations for sustainable textiles: The role of designers and scientists. *Journal of Sustainable Design, 8*(1), 45–62.
15. Quinn, B. (2010). *Textile futures: Fashion, design, and technology*. Berg Publishers.
16. Brown, L. (2023). *Biotechnology in textile design: Lab-grown fibers and their potential*. GreenTech Publishing.
17. Kendall, C. (2019). *Textile design in the digital age*. Thames & Hudson.
18. Green, T., & Knight, P. (2017). Innovations in digital textile printing. *Textile Science Review, 22*(4), 89–105.
19. Martinez, A. (2024). *Eco-textiles: New materials for a sustainable future*. GreenWeave Press.
20. Evans, A. (2015). *Textiles and technology: Pioneering innovations in fabric design*. Bloomsbury Publishing.
21. Farrer, J., & Harris, M. (2017). *Sustainable textiles: Lifecycles and new materials*. Woodhead Publishing.
22. Sparke, P. (2008). *The genius of design*. Quadrille Publishing.
23. Johnson, K., & Lee, S. (2021). AI in fashion: Transforming the design-to-consumer pipeline. *Technology and Fashion Review, 19*(3), 10–25.
24. Green, A., & Taylor, R. (2021). Sustainable materials in the fashion industry: Challenges and solutions. *Journal of Environmental Design, 34*(2), 45–62.
25. Jones, P., & Brown, E. (2020). Natural dye techniques in contemporary fashion design. *Fashion and Culture, 12*(4), 67–80.
26. White, S. (2020). Digital printing in sustainable fashion: Techniques and impacts. *Design and Sustainability, 10*(5), 22–38.
27. Miller, C., & Davis, H. (2021). Transparency and trust in sustainable fashion: Consumer perspectives. *Ethical Fashion Studies, 8*(1), 23–40.
28. Gwilt, A. (2014). *A practical guide to sustainable fashion*. Bloomsbury Publishing.
29. Singh, P. (2024). *Mushroom leather and beyond: Exploring bio-based textiles*. BioFashion Press.
30. Moore, S. (2022). *Sustainability in textile design: Methods and impact*. EcoFabrics Publishing.
31. Smith, J. (2021). Advancements in adaptive textiles: Blending fashion and functionality. *Wearable Technology Journal, 15*(2), 33–49.
32. Pritchard, A. (2021). *Creative processes in the fashion industry*. Routledge.
33. Parker, L. (2020). The impact of technology in textile design. *Journal of Textile Innovation, 19*(3), 56–67.
34. Fletcher, K., & Grose, L. (2012). *Fashion and sustainability: Design for change*. Laurence King Publishing.
35. Watson, E. (2024). *Algae-based textiles: A revolution in sustainable fabric design*. Future Threads Publishing.
36. Bowles, M., & Isaac, A. (2012). *Textile design: Principles, advances, and applications*. Woodhead Publishing.

37. Gordon, S., & Hill, C. (2019). Sustainability challenges in textile design: A review. *Journal of Textile Design, 26*(3), 209–225.
38. Clazie, A. (2013). Building your textile design portfolio: A step-by-step guide. *Textile Design Journal, 45*(2), 88–102.
39. Linton, J. (2012). Building a portfolio for textile designers. *The Textile Design Review, 3*(1), 49–54.
40. Muthu, S. S. (2014). Assessing the environmental impact of textiles and the clothing supply chain. Woodhead Publishing.
41. Tharp, T. (2003). *The creative habit: Learn it and use it for life*. Simon & Schuster.

Emerging Trends and Forecasting in Textile Design

Babita Bhandari and Monika Negi

1 Introduction

As per the report by Mordor Intelligence, the global market size of the textile industry is USD 748 billion in 2024, increasing at a CAGR of 3.52%. During the forecasted period of 2024–2029, it is assumed to reach USD 889.24 billion by 2029. It suggests significant growth, primarily driven by demand for apparel which is influenced by the expansion of e-commerce platforms and the fashion industry. The textile industry is divided into segments based on raw materials like natural and synthetic fibres. The application of textiles spans across fashion, household, and technical sectors, with fashion being a dominant segment [1].

Textile design involves creating innovative textile products that amalgamate aesthetics and usability. The environment, encompassing lifestyle, weather, culture, and politics, serves as a source of inspiration for designers. Together, these factors influence creativity, which leads to various interpretations, original ideas, and inventive designs in various settings. Value addition could be done through various techniques like dyeing, printing, painting, embroidery, finishing, and other embellishments, etc. [2].

The Textile design industry is dynamic and it is crucial to stay relevant in the market by producing trendy textile products with aesthetics. Nowadays, textile designs beautifully amalgamate technological advancements and sustainability aspects and at the same time consider sociocultural narratives with the use of

B. Bhandari
National Institute of Fashion Technology, Kangra, Himachal Pradesh, India
e-mail: babita.bhandari@nift.ac.in

M. Negi (✉)
Department of Fashion Design, Dev Bhoomi Uttarakhand University,
Dehradun, Uttarakhand, India

Sadhna et al. (eds.), *The Art and Craft of Modern Textile Design*, Sustainable
Textiles: Production, Processing, Manufacturing & Chemistry,
https://doi.org/10.1007/978-3-031-86797-2_11

traditional textile techniques. All these factors are taken into consideration by the designers along with ethical aspects. The landscape of the textile industry is constantly changing as textile designers are challenged to design new creative and innovative products while addressing the global market within a specified cost limit [3].

2 Forecasting Future Design Directions

The textile design field is always governed by trends and advancements which is only possible if textile designers are well acquainted with the upcoming trends and what consumers will be demanding in the near future. The market-relevant product can only be designed if one has fair understanding of what colours, what fabrics, and what aesthetics are going to be liked and consumed by prospective consumers in the industry. Hence forecasting and predicting those major trends become very crucial as these act as a major direction that determines the future course of action for many stakeholders [4]. Trend forecasting is the direction in which a particular phenomenon is going to evolve which captures the essence and mood of the times. The trend service companies include WGSN, Trendstop, Trendzoom and Fashion Snoops, etc. After studying and analyzing various key factors like consumer behaviour and preferences, market dynamics and emerging trends, different forecast service companies predict textile trends. All these predictions give a direction for designing products and plan better inventory and positioning in the marketplace. Different trend-forecasting magazines and reports are developed by the agencies which are further referred by the various brands and companies at different levels. The forecasters are alert and observant of the cultural changes, socio-economic factors, and different innovative advancements in different industries [5]. These agencies predict the upcoming forecasts in terms of themes, colours, fabrics, textures, prints, patterns, trims, and silhouettes at least 2 years in advance which influences and governs the design process remarkably for different textile companies [6, 7]. Due to the fast-changing nature of trends in fashion, forecasters need to change and modify the methods and techniques of forecasting, wherein agencies have been following various advanced methods like neural networks, fuzzy logic, data mining etc. These technologies help in providing an accurate picture of forecast which the textile companies can use to reduce understocking or overstocking hence significantly reducing waste generation and optimizing sales. Additionally, the integration of technology and e-commerce has compressed the supply chain, necessitating forecasters to consider the global market systems and the online shopping mechanisms that have shaped consumer behaviour [5, 8–10].

The forecasters provide future insights that inspire and inform decision-makers. They help retailers and brands to make the right choices for their markets and increase sales. The forecasters also consider the sustainability issues, as seen in the shift towards low-carbon materials and digital technology's impact on design elements [11, 12]. There is a great relationship between forecasting and design wherein forecasting informs and influences design process and design influences the forecasting in a continuous cycle [13, 14].

Forecasters need to work with different stakeholders of the textile supply chain following a timeline starting from a few months and extending to 10 years in advance of product manufacturing as explained in Fig. 1. Each type of forecast provides insights into the right selection of fabrics, colours and future design directions etc. There are two types of forecasting done in the textile industry.

2.1 *Short-Term Forecasting*

In short term forecasting, the microtrends don't last more than a season. This type of forecasting shows how the collection's color and style change over time. Most apparel companies subscribe to one or more service, whose purpose is to survey the market and report on developments in colour, textiles, and style directions.

2.2 *Long-Term Forecasting*

Long-term forecasts are determined based on changes in demographics, patterns of textile production and marketing, and, more importantly, changes in the lifestyle of consumers. This kind of forecast should still be valid for at least several years [5].

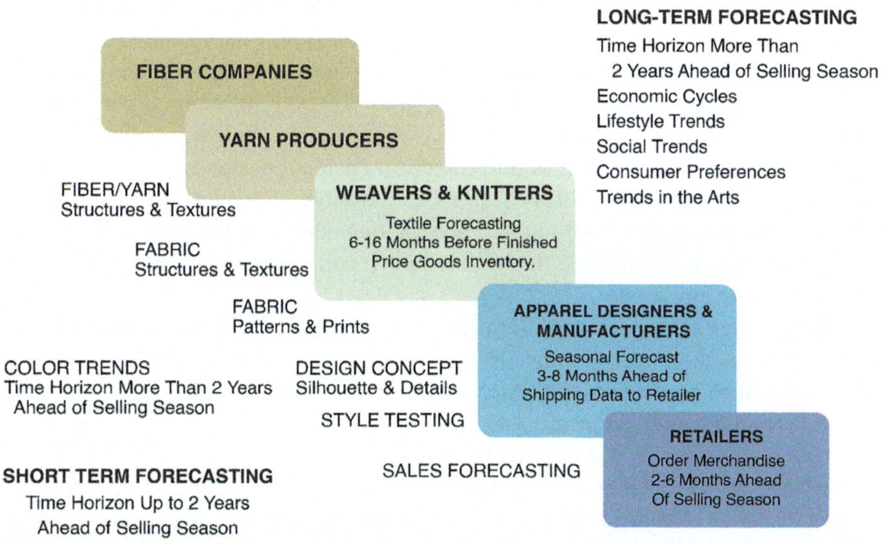

Fig. 1 Short- and long-term forecasting timeline within the manufacturing cycle [5]

The textile design process, encompasses designing patterns and textures on textiles using weaving, printing, and dyeing, which requires a combination of artistic creativity and technical expertise to produce materials that are distinct, functional, and beautiful all at the same time. The choice for colors, type, and scale of motifs, and type of fabrics are decided based on demographic attributes. Other factors that affect the textile design trends include cultural aspects, social values and beliefs, social media trends, developments in fashion, interior and visual arts etc.

Hence studying demographic trends along with other major aspects is crucial for accurate forecasting of textile design trends for better market reach [15].

3 Recent Trend-Setting Designs

Textile designers should try to bring novelty and creativity to their products for better product differentiation. Designers are keeping up with the latest trends and technologies and striving to stay ahead. Some recent trendsetting designs and technology are described below:

3.1 3D Printed Designs

Many innovative textile companies are enabling designers to create interesting visual and tactile effects on diverse range of materials. Various interesting tactile patterns and objects can be directly created onto the fabric surface like denim, cotton, polyester, and linen with the help of 3D printing technology. Designing using 3D printing technology is evolving as the new trend since the technology is evolving. Various technology companies are launching new design technologies like Stratasys' 3DFashion technology creating upto 600,000 different colour shades of varying transparency levels with a range of textures and finishes. Direct-to-textile 3D printing converts 2D designs into three-dimensional designs rendering new life into fabrics and patterns. Endless designs are possible with customized and unique personalized designs on clothing and accessories adding 3D printed elements. 3D printed designs also reinforce the sustainability parameter to product designs as it does not create any waste.

A new range of eco-friendly 3D printed cushions were created using Stratasys' 3DFashion technology by using discarded scrap fabrics from textile companies across the globe by the company Atelier des Refusés. Creative textures and aesthetics were added using 3D printing technology adding exquisite 3D printed embellishments with reduced waste generation [16].

Researchers and designers are exploring the use of 3D printing to produce not only wearable technology and accessories but also everyday clothing, although this application is less common [17]. It offers a new dimension of creativity and functionality in fashion and textile industries, allowing for the production of textiles with varying thickness, fibers, and filaments [18].

3.2 *Interactive Textile Design for Emotion Regulation*

Interactive smart textiles for emotion regulation represents a trend-setting design due to their innovative integration of technology and textiles to address emotional well-being. These textiles incorporate sensors and actuators that respond to physiological signals, enabling the wearer to regulate emotions through movement-based interactions and biofeedback mechanisms. The design of such textiles is informed by the 'Research through Design' approach, which emphasizes the iterative development of prototypes to enhance user experience [19]. Interestingly, while the trend is gaining traction in various fields, including healthcare, where smart textiles are used to mitigate stress and improve the emotional state of healthcare workers, it also intersects with the broader Internet of Things landscape in fashion innovation. The trend is propelled by the desire to blend technical functionality with emotional expression, leading to the development of smart clothing that not only satisfies technical requirements but also supports human emotional needs. It indicates a broader movement towards integrating technology into everyday life in a manner that is both functional and emotionally resonant [20, 21].

3.3 *Circular Textile Design*

At present there is a great awareness of the actions of designers concerning the responsibility for people and planet. The principles of the circular economy are characterized by the sustainable management of resources through the reduction, reuse, and recycling of materials. The textiles are designed in such a manner that all other materials used in the product can used back at different stages of the textile manufacturing cycle through various means, hence it does not form a linear chain of production. So, materials should be selected in such a manner that after each lifespan of the product, the material could be easily upcycled, downcycled or recycled for a new lifecycle [22]. Interestingly, while the circular economy is a promising direction for sustainability, it presents challenges such as the need for systemic change and the adoption of new business models.

Upcycling has been commonly used sustainable and creative means that can contribute to the circular economy by adding value to the products. Moreover, trends of clothing reuse and recycle has been emerging on a community level, hence highlighting the social aspect of sustainable fashion adoption. Circular economy in textiles is being considered an essential step in ensuring a sustainable future rather than a mere trend [23, 24].

3.4 Smart Textile Design

Smart textiles are the integration of advanced materials and technologies that enable textiles to respond to external stimuli at various levels. Smart textiles, often called electronic textiles or e-textiles, incorporate functionalities such as chromic changes, energy harvesting, and sensing capabilities, which have applications across various sectors including healthcare, military, and consumer electronics [25]. Interestingly, while there is significant progress in the field, challenges remain in terms of performance characterization, scalability of production methods, and integration of electronic components into textiles. Additionally, the development of textiles with phase change materials (PCMs) for thermoregulation showcases the multifunctional potential of these materials. Despite the advancements, ongoing research is addressing the challenges related to integration, production scalability, and performance to fully realize the potential of smart textiles in various domains [26, 27]. Figure 2 mentions examples of smart textiles like smart yoga pants, hug shirt, sound shirt, shocks with sensors, Levi's Jacket with sensors and Mercury jacket.

4 Major Trends in the Textile Industry

The amalgamation of technological advancements with innovative approaches in design and production represents a paradigm shift. Design-driven innovation is being adopted in the textile industry to explore the huge potential and scope of textile design. Some major trends in textile design are discussed below:

4.1 Technological Innovations

Technological progress in textile design has had a significant impact on both the design process and the finished products. Innovations such as CAD software, digital printing processes, and intelligent textiles have enabled designers to create elaborate patterns, personalized textiles, and functional materials with superior performance qualities [28].

Technological innovations also promote the sustainable practices in the textile industry through less waste and energy consumption. Moreover, fabrics with enhanced properties like moisture management, UV protection, antimicrobial properties while also considering the style and functionalities as per changing consumer preferences [29].

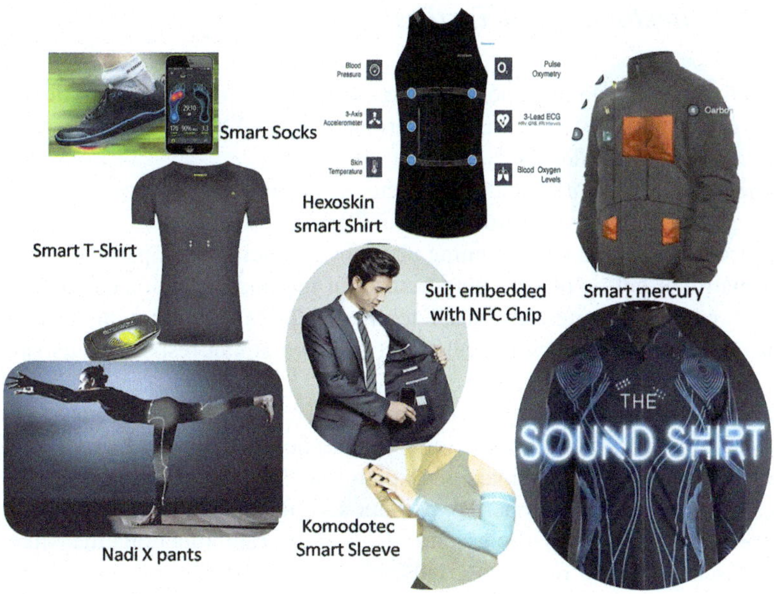

Fig. 2 Examples of wearable electronic smart textiles [27]

4.2 *Smart Fabrics*

Smart textiles integrate advanced technologies to offer functionalities beyond traditional fabrics. Smart textiles are called smart as these can perform very specialized activities as these fabrics can change their color, temperature, behaviour, based on the requirement of the wearer. Also these textiles can protect the user from hazards of various degrees, enhance the performance of the wearer etc. These digital components, batteries, sensors, or electronic chips will be integrated into garment production processes [30].

These developments include temperature-regulating clothing and wearable health monitoring, as well as new ways for users to interact with garments. Some examples include thermochromic fabrics, which change color with temperature, conductive fabrics, which transfer electrical information, and self-healing materials, which fix punctures. Other examples are photochromic textiles that react to light, pressure-sensitive fabrics for medical and interactive applications, shape memory textiles for aerospace and fashion, and water-resistant, breathable materials for outdoor use. LED-illuminated textiles, temperature-regulating materials in activewear, and antimicrobial treatments in medical and activewear all contribute to the consolidation of these diverse uses [31].

4.3 Nanotechnology in Textiles

Nanotechnology is bringing about a microscopic revolution in textiles. The increased desire for sustainable, long-lasting garments is making it possible to include nanoparticles into textiles 24.

Nanotechnology works with nanometer-scale materials, whose characteristics change dramatically at small dimensions [32].

Electrospinning and split spinning nanofibers generate nanoparticles that can be coated with sol-gel or plasma polymerization methods. Nanometer-scale components of nanocomposites increased textile qualities, resulting in an electrical, mechanical, optical, and biological smart textile that can respond to its surroundings [33].

Nanomaterials promise a great future; they can be utilized in making fabrics for particular functionalities that can sense and respond to different kinds of stimuli, electrical or physiological signals. Incorporation of nanoparticles or nanofibers into fabrics enhances several properties without significantly increasing weight, thickness, or stiffness. Nanoparticles can also incorporate functionalities like static charge discharge, stain resistance, and conductivity into fibers without losing care for comfort and drape of the final textile [32].

Nano-fibers possess excellent properties, such as excellent filtration and high permeability with thin layers. They can reject poisonous gases, bacteria, viruses, and other harmful airborne contaminants. This makes them very versatile and useful [33].

Nanotechnology offers new features that include energy storage and communication through clothing. This includes stain or wrinkle-resistant threads, temperature responsive warmers, and odor-resistant socks. Nano-enhanced textiles impart properties such as water/soil repellence, anti-bacterial, UV protection, and self-cleaning abilities [34].

4.4 Artificial Intelligence Design

Artificial Intelligence (AI) has been used in textile design from the stage of predicting trends.These are also used in different stages of production and quality control. AI gives flight to the designers' imagination by providing endless design possibilities in terms ofcolour palette, texture, surface techniques, fabric materials, and finishes till final garment ideation in short span. Artificial Intelligence and Machine Learning application has improved supply chain management, market trend forecasting, and manufacturing efficiency. AI enhances the creativity of designers as it can do mundane tasks, allowing designers to focus on generating ideas. It acts as a virtual assistant by performing minor details so that the designers can focus on their creative processes [35].

4.5 *Biodegradable Textiles*

Bio-degradable textile materials are emerging as a result of societal concern over the emission of textile waste. The creation of textiles that decompose organically and reduce environmental harm has improved with the advancement of material science. Organic cotton, plant-based textiles, and the circular fashion concept are among the initiatives. Reducing, reusing, and recycling are important in the fashion business since the industry is currently shifting to sustainable materials that minimize waste while offering a more environmentally friendly cycle of life for clothing [36].

4.6 *Virtual Try-On Technologies*

Those working in the textile and garment industries can now see and experience things virtually thanks to this technology. People may make decisions to buy the products from the comfort of their homes thanks to the immersive 360-degree experience that is made possible by AR and a high-tech visor. Therefore, the state-of-the-art of this technology guarantees that the decision-making process is optimized and the purchasing experience is improved by altering the way we search for and obtain our goods [37].

Virtual Try-On, or VTO, technology has, in a sense, changed what consumers purchase; they will now have a relationship with fashion that has never been seen before. One can digitally wear elegant sunglasses, put on stylish sneakers, accessorize with elegant jewelry, apply seductive makeup, or even try on a stunning formal dress with a few clicks on a mobile application or website—all from the comfort of their own home. Since the majority of the dressing rooms at these chaotic physical retail establishments are closed, online shopping can flourish. This has been nicely complemented at a time when internet shopping has become extremely popular and VTO, as technology, has opened up new, revolutionary possibilities for how customers may engage and play with fashion items [38].

4.7 *Self-Cleaning Textiles*

Sustainable textile innovations also focus on the attributes of final products. A new advancement is the development of self-cleaning garments, which incorporate tiny copper or silver structures within the fibers. When exposed to sunlight, these nanostructures absorb solar energy, energizing the metal atoms to break down dirt on the fabric [39].

4.8 Rechargeable Fabrics

The future of fashion is in smart clothing that interacts with electronics. Currently, these garments require energy from batteries or charging systems, but soon they may recharge themselves. Researchers at Georgia Tech have developed yarns that harvest energy from static electricity generated by friction between different materials. When integrated into fabrics like socks, sweaters, and pants, they can capture energy from body movements, benefiting both the environment and our health [40].

5 Breakthrough Experimentations in Textile Design

5.1 Bio-Couture

Bio couture, a term that encompasses the use of biological processes in the creation of textiles, plays a pivotal role in the evolution textile design. Concurrently, the pursuit of sustainability in fashion has led to the exploration of bio-cellulose and other biomaterials that are notonly sustainable but also biodegradable and reusable, thus redefining the aesthetics and functionality of textiles. Biocouture sits at the intersection of three distinct disciplines: fashion design, biotechnology, and material engineering [41]. It is a new concept of making high-end clothes where cloths are designed and produced by living microorganisms such as bacteria, fungi etc. in the presence of other raw materials like starch, woodpulp, sugar etc. under highly optimized conditions [42, 43]. Figure 3 shows the bio-garment produced by designer Suzanne Lee.

5.2 Animated Textiles

It is a new area of textile design where an interactive system is created in textile itself. It renders animated properties to textiles through variousphysical,biologicaland computational efforts [44]. This concept explores the idea of woven textile development that can change in the form of textures and shapes in the presence of heat. However, animated textiles have a great potential as these can exhibit responsive behaviors through their inherent material properties. The research in animated textiles spans from the development of responsive fabric structures to the incorporation of intelligent materials for health monitoring and emotional regulation etc. There needs to be a paradigm shift towards recognizing and appreciating textiles as active participants in design, rather than passive elements. This change in perspective calls for a more comprehensive understanding of textiles in interactive systems and demands the extensive cross-disciplinary collaboration in the field [45].

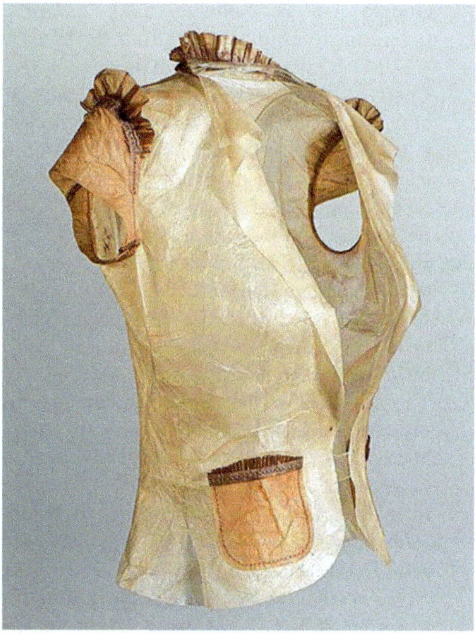

Fig. 3 Suzanne Lee, bio-couture in microbial cellulose, 2010 [43]

6 Conclusion

In the era of ever-evolving design landscape, it is very crucial to predict trends. The forecasting helps the design fraternity to direct their efforts along the lines of forecasted trends. Having a relevant idea about how trend forecasting takes place, and how it affects the textile industry is very important for any designer to produce consumer-relevant products. Further having an idea about trends be it related to materials, techniques, technology or end uses, is equally important.Prominent forecasting agencies such as WGSN and Promostyl offer comprehensive guidance on industry trends and predictions. As environmental consciousness grows, designers are increasinglyadopting eco-friendly materials and methods, leading the way towards a more sustainable fashion sector. Cutting-edge developments like smart textiles, nanotechnology, AR, VR, bio-couture and 3D printing are transforming fabric creation and application, opening up new avenues for personalization and functionality. The textile industry is experiencing expanded horizons due to trend-setting designs and emerging concepts in circularity, interactive textile design, smart textiles, technological advancements, and AI.

References

1. Market Research Company. (n.d.). *Textile industry: Analysis, market overview, report & growth.* https://www.mordorintelligence.com/industry-reports/global-textile-industry%2D%2D-growth-trends-and-forecast-2019%2D%2D-2024

2. Google Arts & Culture. (n.d.). *Indian textiles: Nature & making*. https://artsandculture.google.com/story/indian-textiles-nature-making-victoria-and-albert-museum/YAUROUQuPAXOJw?hl=en
3. Cherryliu. (2015). *Post A: Cultural influences on textile design*. https://indonesiadesignstudio.blog/2015/05/01/post-a-cultural-influences-on-textile-design/
4. Klosowski, A., & Poland, W. (2023). Condition analysis and forecasting in the fashion industry: A comprehensive study. *International Journal of Economics Finance & Management Science, 8*(7), 16–19. https://doi.org/10.55640/ijefms-9132
5. Divita, L. (2019). *Fashion forecasting* (5th ed.). Fairchild Books.
6. King, J. A. (2011). Colour trend forecasting and its influence on the fashion and textile industry. In *Textile design* (pp. 193–206). Elsevier. https://doi.org/10.1533/9780857092564.3.193
7. King, J. A. (2014). Colour trend forecasting and its influence on the fashion and textile industry. In *Textile design* (pp. 193–206). Elsevier. https://doi.org/10.1016/b978-1-84569-646-7.50009-3
8. Thomassey, S. (2010). Sales forecasts in clothing industry: The key success factor of the supply chain management. *International Journal of Production Economics, 128*(2), 470–483. https://doi.org/10.1016/j.ijpe.2010.07.018
9. Gladys, A., & Olalekan, A. (2021). A machine learning model for predicting colour trends in the textile fashion industry in south-west Nigeria. *International Journal on Integrated Education, 4*(2), 174–188.
10. Jain, D. M. (2022). SWOT analysis of e-fashion & textile industry. *International Journal of Home Science, 8*(3), 288–292. https://doi.org/10.22271/23957476.2022.v8.i3e.1385
11. Liu, Y. Y., & Ma, D. L. (2013). The performance and development trends of low-carbon textile materials. *Applied Mechanics and Materials, 291–294*, 1421–1424. https://doi.org/10.4028/www.scientific.net/amm.291-294.1421
12. Lee, N., & Suh, S. (2024). How does digital technology inspire global fashion design trends? Big data analysis on design elements. *Applied Sciences, 14*(13), 5693. https://doi.org/10.3390/app14135693
13. Chuprina, N. V., Paranko, N. P., Kovalchuk, M. O., Prasol, S. I., Baranova, A. I., & Omelchenko, H. V. (2021). Style direction "unisex" as a design practice in fashion design. *Art and Design, 1*, 9–20. https://doi.org/10.30857/2617-0272.2021.1.1
14. Ujiie, H. (2002). Textile design education in digital inkjet fabric printing. *NIP & Digital Fabrication Conference, 18*(1), 254–257. https://doi.org/10.2352/issn.2169-4451.2002.18.1.art00063_1
15. Anonymous. (2024). *Socio-cultural and environmental influences on design*. https://studyrocket.co.uk/revision/a-level-textiles-ccea/module-design-process-and-product-development/socio-cultural-and-environmental-influences-on-design
16. Anon. (2023). *The new era of sustainable luxury*. https://www.stratasys.com/en/resources/case-studies/atelier-des-refuses-3d-printed-cushions/
17. Jadhav, A., & Jadhav, V. S. (2022). A review on 3D printing: An additive manufacturing technology. *Materials Today: Proceedings, 62*, 2094–2099. https://doi.org/10.1016/j.matpr.2022.02.558
18. Dehghani, M., & Goyal, P. (2022). Design and development of textile fabrics using 3D printing technology. *ECS Transactions, 107*(1), 19313–19323. https://doi.org/10.1149/10701.19313ecst
19. Jiang, M., Bhömer, M. T., & Liang, H. N. (2020). Exploring the Design of Interactive Smart Textiles for emotion regulation. In C. Stephanidis, V. G. Duffy, N. Streitz, S. Konomi, & H. Krömker (Eds.), *HCI international 2020—Late breaking papers: Digital human modeling and ergonomics, mobility and intelligent environments. HCII 2020* (Lecture notes in computer science) (Vol. 12429). Springer. https://doi.org/10.1007/978-3-030-59987-4_22
20. Jiang, M., Wang, Y., Nanjappan, V., et al. (2024). E-textiles for emotion interaction: A scoping review of trends and opportunities. *Personal and Ubiquitous Computing, 28*(3–4), 547–577. https://doi.org/10.1007/s00779-024-01793-w

21. Jiang, M. (2021). Movement-based interactive textiles design for emotion regulation. *The Design Journal, 24*(3), 477–488. https://doi.org/10.1080/14606925.2021.1903678
22. Unal, F., Avinc, O., & Yavas, A. (2020). Sustainable textile designs made from renewable biodegradable sustainable natural abaca fibers. In *Sustainability in the textile and apparel industries* (pp. 1–30). Springer. https://doi.org/10.1007/978-3-030-37929-2_1
23. Sowmya, A., & Nag, S. (2024). From textile waste to fashion: A study on reusing textile waste for promoting sustainable fashion for its prospects. *International Journal for Research in Applied Science and Engineering Technology, 12*(2), 970–976. https://doi.org/10.22214/ijraset.2024.58492
24. Avadanei, M., Souto, R., Belakova, D., Martinez, R. O., Mouazan, E., & Sivevska, N. (2021). *E-learning platform of eco-design in textile and fashion sectors towards a circular textile.* https://doi.org/10.12753/2066-026x-21-150
25. Júnior, H. L. O., Neves, R. M., Monticeli, F. M., & Dall Agnol, L. (2022). Smart fabric textiles: Recent advances and challenges. *Textiles, 2*(4), 582–605. https://doi.org/10.3390/textiles2040034
26. Dutta, S., & Bansal, P. (2024). Evolution of sustainable wearables: Integrating cutting-edge techniques for future textile innovation. *Research Journal of Textile and Apparel.* https://doi.org/10.1108/rjta-03-2024-0044
27. Takebira, U. M., Mehbub, A., Mondal, M. I. H., Islam, M. M., Mohibullah, A. T. M., & Pervez, M. N. (2024). Wearable smart textiles: Recent development and applications. In *Smart textiles from natural resources* (pp. 395–442). Elsevier. https://doi.org/10.1016/b978-0-443-15471-3.00003-0
28. Chen, M., & Wang, L. (2015). Fashion design innovation based on computer aided design and electronic technology. In *International Conference on Engineering and Technology Innovation* (pp. 43–47).
29. Lauria, A. (2017). *Technological advancements in textiles.* Textile Today. https://www.textiletoday.com.bd/technological-advancements-textiles/
30. Hartin, T., & Muscato, C. (2023). *Smart textiles: Applications & examples.* https://study.com/academy/lesson/what-are-smart-textiles.html
31. Meena, J. S., Choi, S. B., Jung, S. B. J., & Kim, J. W. (2023). Electronic textiles: New age of wearable technology for healthcare and fitness solutions. *Materials Today Bio, 19,* 100565. https://doi.org/10.1016/j.mtbio.2023.100565
32. Ali, K., Yetisen, A. K., Qu, H., Manbachi, A., Butt, H., Dokmeci, M. R., Hinestroza, J. P., Skorobogatiy, M., Khademhosseini, A., & Hyun Yun, S. (2016). Nanotechnology in textiles. *ACS Nano, 10*(3), 3042–3068. https://doi.org/10.1021/acsnano.5b08176)
33. Kiron, M. I. (2013). *Application of nanotechnology in textile industry.* https://www.fibre2fashion.com/industry-article/7135/application-of-nanotechnology-in-textile-industry
34. Bose, P. (2020). *The role of nanotechnology in the production of fabrics.* https://www.azonano.com/article.aspx?ArticleID=5501
35. Anwer, H., Ali, M., & Jamshaid, H. (2024). Applications of artificial intelligence in textiles and fashion. In H. Jamshaid & A. Dad (Eds.), *Creative textile industry. SDGs and textiles.* Springer. https://doi.org/10.1007/978-981-97-3802-1_8
36. Nofal, R. M. (2022). Biodegradable textiles, recycling, and sustainability achievement. In *Handbook of biodegradable materials.* Springer. https://doi.org/10.1007/978-3-030-83783-9_54-1
37. Tailoor. (2023). *Virtual try-on in fashion: What is it and what are its benefits?* https://tailoor.com/virtual-try-on-in-fashion-what-is-it-and-what-are-its-benefits/
38. Lal, K. (2023). *Does virtual try-on technology actually work for fashion?* https://fashionista.com/2023/10/virtual-try-on-apps-technology-clothing-accessories
39. Afroz, S., Azady, M. A. R., Akter, Y., Ragib, A. A., Hasan, Z., Rahaman, M. S., & Islam, J. M. M. (2021). Self-cleaning textiles: Structure, fabrication and applications. In M. I. H. Mondal (Ed.), *The textile institute book series, fundamentals of natural fibres and textiles* (pp. 557–597). Woodhead Publishing.

40. Zhang, N., Huang, F., Zhao, S., Lv, X., Zhou, Y., Xiang, S., et al. (2020). Photo-rechargeable fabrics as sustainable and robust power sources for wearable bioelectronics. *Matter, 2*(5), 1260–1269. https://doi.org/10.1016/j.matt.2020.01.022
41. Atmos. (n.d.). *The Designer Growing BiocoutureFrom Lichens and Slime Mold*. Atmos. https://atmos.earth/the-designer-growing-biocouture-from-lichen-and-slime-mold/
42. Kwak, A. (2022). *Evolution of biocouture 2022: The Garnette report*. The Garnette Report. https://thegarnettereport.com/fashion/evolution-of-biocouture-2022/
43. Stear, N. (2014). *BioCouture: Growing a wardrobe from bacteria | live eco*. Live Eco. https://liveeco.co.za/2014/05/19/biocouture-growing-wardrobe-bacteria/
44. Buso, A., McQuillan, H., Jansen, K., & Karana, E. (2022). The unfolding of textileness in animated textiles: An exploration of woven textile-forms. In D. Lockton, S. Lenzi, P. Hekkert, A. Oak, J. Sádaba, & P. Lloyd (Eds.), *DRS2022: Bilbao, 25 June–3 July, Bilbao, Spain*. https://doi.org/10.21606/DRS.2022.612
45. Luprano, J. (2012). Progress in interactive textiles for health monitoring. In *Advances in Science and Technology* (Vol. 86, pp. 1–8). https://doi.org/10.4028/www.scientific.net/ast.86.1

Interactive Multimedia in Textile Design

C. Anisha and M. Nandhini

1 Introduction

The method of designing patterns for weaved, knitted, printed, or surface-adorned textiles is referred to as textile designing. It modifies how textiles and other materials look. Garment design, clothing design, and various techniques for producing textiles are all included in the creative realm of designing textiles. Interior and fashion designers collaborate closely with textile designers [1]. The capacity to create an idea for a certain type of cloth is a must for the textile designer. Furthermore, they must be capable of creating textile designs that are appropriate for a specific use, which is typically on the physique or a specific area. As a way to keep their creations up to date and suitable for the intended use, artists also need to have a solid grasp of colour theory, the latest fashions, and modern design problems [2]. They all cover a wide spectrum of design methodologies and procedures. Creators of printed textiles mostly work with the fabric's surface. Creators of woven and knitted textiles typically start from zero when creating a fabric, choosing yarns and fibres first. In order to create textiles, textile artisans also frequently use other methods. Though it can also be used independently as a skill, the combination of media is frequently used when combined with knitting, weaving, or printing [3].

C. Anisha (✉) · M. Nandhini
Department of Costume and Fashion Designing, Nirmala College of Arts and Science, Thrissur, Kerala, India

Department of Costume Design and Fashion, Nehru Arts and Science College, Coimbatore, India

Sadhna et al. (eds.), *The Art and Craft of Modern Textile Design*, Sustainable Textiles: Production, Processing, Manufacturing & Chemistry,
https://doi.org/10.1007/978-3-031-86797-2_12

2 Application of Tools and Software for Textile Design

Creating, editing, and sharing the fabric's concepts, designs, and colours with other designers in the fashion industry can be facilitated by fabric design tools and software. One may improve overall imaginative abilities and interpersonal skills by using these online tools to work together on an assignment, obstacle, or accumulation. This post will teach the best ways to work together with other designers, as shown in Fig. 1, who create clothes using some of the most well-liked and practical textile design software and tools [4].

Selecting a fabric creation programme that most closely matches the demands and tastes is necessary when working alongside other designers in the fashion industry. The choice must be made after considering a number of factors, including the program's capabilities and processes, interoperability and cooperation, convenience to use, expenses, and accessibility. Adobe Photoshop and Illustrator, Procreate, and Pattern Design are some of the often-used software and tools for textile design [1].

The exchange of textile design files is a subsequent stage in working with different fashion illustrators. There are numerous ways one may distribute the data based on the application and application that use. An incredibly straightforward and widely used method of sharing documents is via messages, although cloud computing platforms like Dropbox, iCloud, and Adobe Creative Cloud provide greater safety and ease. Online marketplaces like Spoonflower, Fabric on Demand, or Pattern Design let you publish and display your ideas in addition to finding and connecting with different creators enabling an additional inventive and participatory approach. To organize the relationship and find inspiration from work by others, you may also make use of tools like folders, comments, version history, galleries, forums, and contests [5].

Reviewing and modifying the textile layout files is usually the third phase of collaborating alongside different fashion designers. There are variety of method for

Fig. 1 Tools and software for textile design [4]

reviewing and editing the data based on the programme and application that use. One can utilise the programme or tool's preview option, for instance, to examine how the designs seem on various materials, dimensions, and colour schemes. Anyone may also pose and get queries, comments, and ideas regarding your designs using a review option or function inside the tool, email, cloud service, or programme. Furthermore, anyone can enhance or modify the layouts by using an edit option or feature within the app or tool. Lastly, people may team up and contribute alongside additional designers by using an editing tool or feature of the cloud or internet-based service. One can collaborate with fashion professionals to examine and update their textile design files using any of these techniques [6].

The fourth phase of working with different fashion designers is to evaluate and improve the fabric design files. There are several ways to evaluate and improve the content of your files based on the programme and instrument you use. To get an exact idea of the drawings seem in everyday life, one may press them into paper or cloth, for example. As a substitute, people can view whether the designs seem on various materials, outfits, or mannequins by using a replicate tool or feature. Additionally, anyone can evaluate the technical details of the designs you created or examine the effectiveness and widespread appeal of our creations on the World Wide Web by using an analysis option or capability [4].

Finishing up and exporting your textile art documents is the fifth stage of working alongside different fashion designers. There are multiple method of finish and transfer the files based on the programme and application that use. For instance, it's crucial to save the artwork in the standards and style they desire, but it's also more flexible to export them in PNG, JPG, PDF, or SVG, among other file types. A fun approach to present the creations to the public is to post the designs on social media, a website, or a portfolio site. In order to sell or grant licences to present work to clients or customers, one can also publish them online [6].

Savour and commemorate the textile design partnership as the last and sixth step in working with various designers in fashion. There are many various ways to accomplish this based on the programme and instrument you use. With a learning feature or feature of your programme, tool, email, cloud, or online service, for example, you may benefit from both your own and others' expertise, failures, and accomplishments. Additionally, you can utilise the encourage function or capability to share both your and other people's designs, ideas, and stories with an inspiring effect. Lastly, the celebration functionality or characteristic of your software, tool, email, cloud, or online service allows you to celebrate both yourself and others with your own and others' accomplishments, feedback, and recognition [7].

3 Type of Software

Software used for creating and saving textile designs using a computer is known as textile design software. In the apparel sector, buyers are given a few favoured designs to make ordering easier. As a result, the customer asserts that changing the textile's colour and design is regularly required. Handmade design is unlikely to be flawless and may require a long time. In this instance, using textile design tools for fashion designers will make the process really quick and effortless [8].

3.1 C-Design Fashion

C-DESIGN Fashion is an application that can be utilised by anybody to create clothing, from creating to modelling to making visuals and advertising with images, as shown in Fig. 2. It participates in every phase of life, from concept to output, in addition to permitting the creation of trend. It offers expert capabilities and refinements, as well as a comprehensive and user-friendly interface. Produce, modify, arrange, and distribute your work. Create, arrange, and distribute your scientific documents among your freelancers. Create demonstrations, publications, and groupings, and model the windows and shop layout. You'll adore how quickly, easily, and adaptably your creative existence develops in the near future. Without a doubt, this textile design programme is perhaps the most sophisticated available. In addition to offering the most recent version of creative tools, C-Design uses precise technical records for a fresh, organized process of development that includes colour matching, labelling, measuring charts, and much more [9].

Fig. 2 C-DESIGN Fashion [9]

3.2 *Planning of Enterprise Resources (ERP)*

The software packages are utilized for major corporate applications and are separated into one section. A computing programme called ERP, or enterprise resource planning, was created especially for use in factories and business premises. The primary goal of ERP is to provide all of an organization's data underneath a single canopy to support decisions and planning with total insight. It is composed of several components that cooperate to oversee the organization's overall exchange of data, preservation, and access. ERP is the computerization of all processes that are in any way engaged in the process of production, from raw materials to the finished product. These individuals ought to practice complete supply chain control to ensure client requests are delivered on schedule. ERP manages the entire clothing manufacturing process in addition to assisting with the item's designing (CAD) and production (CAM) phases [10]. ERP covers nearly every function, including business, client relations, finance, employee relations, manufacturing, shipment, purchases, and business performance and administration. It also lowers expenses and disinformation while enhancing the quality of the products. The availability of information on the internet has increased the calibre of the process of making choices [1].

3.3 *SnapFashun*

SnapFashun is a feature-rich style planning application aimed at small and medium-sized businesses and startups. It is a fabric or fashion-related application that is user-friendly, cost-effective, and highly efficient. SnapFashun offers full offerings made specifically for applications on the web, as given in Fig. 3. This web-based platform for designing clothes provides CAD tools, fashion illustrations, print and cut patterns, presenting tools, and concept exports all in one location. Children studying the textile arts will find it practical, but professionals will find it to be incredibly powerful

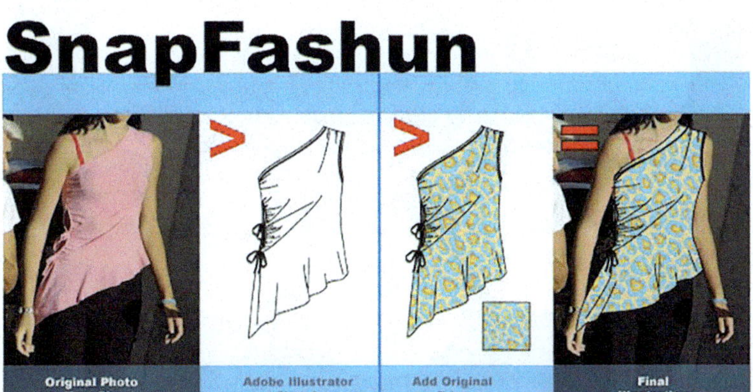

Fig. 3 SnapFashun [11]

technology. Several designers have already expressed interest in this software because of the multitude of pre-design layouts and drawing features [11].

3.4 WinMan

The manufacture and competitiveness of fabric-based goods. Many difficult jobs in the industry can be completed quickly with WinMan, a system that is not flexible. Budgeting and arranging, managing documents, managing projects, logistics administration, CRM, WMS, manufacturing administration, and design configurator, among numerous additional applications, are all possible with it [5].

3.5 TexAMS ERP

TexAms ERP is the world's top supplier of software and products specifically for the garment sector. The TexAMS system for enterprise resource planning serves as a timetable for the apparel industry, handling bookkeeping for salespeople, the financial sector, effectiveness, and material management for fabrics. Additionally, it may be used for textile product preparation and client support is indicated in Fig. 4.

Fig. 4 TexAMS ERP [10]

TexAMS ERP applications also solve a variety of complicated issues in the clothing sub-industry, including yarn dyeing, spinning, sewing thread, weaving, textile fabrics, and so on. Other fields in which this fabric design software is utilised include the economy, production control, maintenance, quality lab management, yarn dyeing, weaving, sales, distribution planning, and finished goods inventory, among other areas [12]. With the assistance of TexAMS, a particularly created ERP that covers the whole fabric manufacturing process, the company can nearly 100% boost efficiency while reducing waste by 50% [10].

3.6 *Apparel Maestro*

Programs allow Apparel Maestro to maintain track of the quantity of textile items and determine the kind, dimensions, and hue of clothing. Using this application, clothing-related actions can be managed. Added to a host of other qualities, this application can set up prepack setup, generate numbering of lots using three parts (one constant and two user-definable options), measure accounts payable, input colours dynamically, as well as purchase and sell items based on size, style, and colour. Consumers of clothing can reduce costs and inefficiency while boosting profitability, streamlining processes and procedures, excelling in organisation, and improving the economy [13].

3.7 *Browzwear*

Browzwear is a famous method of clothing creation as well. It is a well-known software system for 3D fashion design as well as real prototype creation. Browzwear, which is widely recognised for its state-of-the-art technology, allows designers to create realistic three-dimensional representations of apparel, enabling accurate emulation and presentation. The tool supports multiple stages throughout the garment-making process, from generating patterns through online sample manufacture, which speeds up communication and engagement. Browzwear's emphasis on technological innovations has allowed them to become a prominent player in the continuing modernisation of clothing manufacturing routines, as projected in Fig. 5 [4].

3.8 *Optitex*

Another popular method of developing apparel is Optitex. Optitex is an application utilised in automated two-dimensional and three-dimensional textile creation and structure production. Subsequently, it is renowned for providing advanced

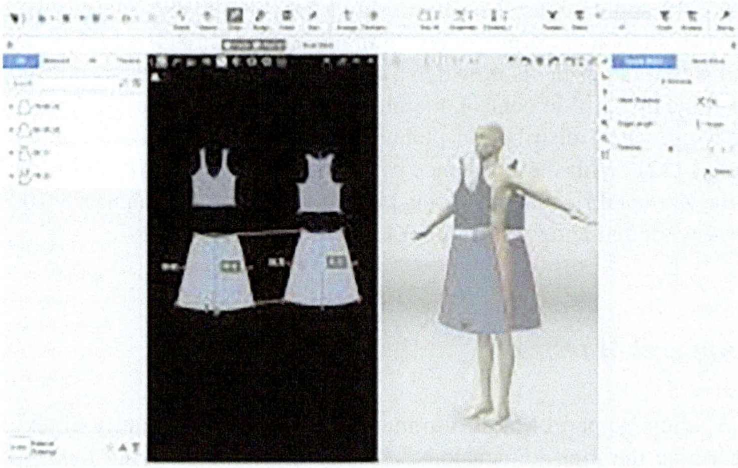

Fig. 5 Browzwear [4]

Fig. 6 Optitex [2]

electronic modelling tools, enabling architects to visualise along with developing garments in three different planes. Optitex provides features such as template generation, assessing, and sign creation, which expedites the garment process of design and enhances accuracy and efficiency in clothing manufacturing, as emphasised in Fig. 6. Widely used in the apparel industry, Optitex aids in the computerisation of both production and styling activities [2].

3.9 TUKA3D

TUKA3D is a state-of-the-art apparel design application specialised in creating and presentation of 3D garments. Known for its innovative equipment, TUKA3D allows architects to create and duplicate realistic 3D representations of apparel, offering a live environment for accurate evaluation and representation. TUKA3D streamlines the techniques of designing and developing designs by providing functions such as pattern creation, draped silhouettes and simulated collection, hence eliminating the need for tangible outputs. TUKA3D, which is widely used in the apparel industry, helps to improve efficiency, shorten time-to-market, and take an ecologically conscious approach to fashion design and production, as shown in Fig. 7 [14].

3.10 Spoonflower's Free Design Tool

Recently, Spoonflower updated its online presence and the methods in which people can view the designs. Users are currently able to see the designs you have projected up in sheets of paper, blinds, mattresses, table linens, and further, rather than just a swatch and a fat quarter. Essential but constrained. Excellent for novices. Uploads straight to Spoonflower for sample orders in Fig. 8 [13].

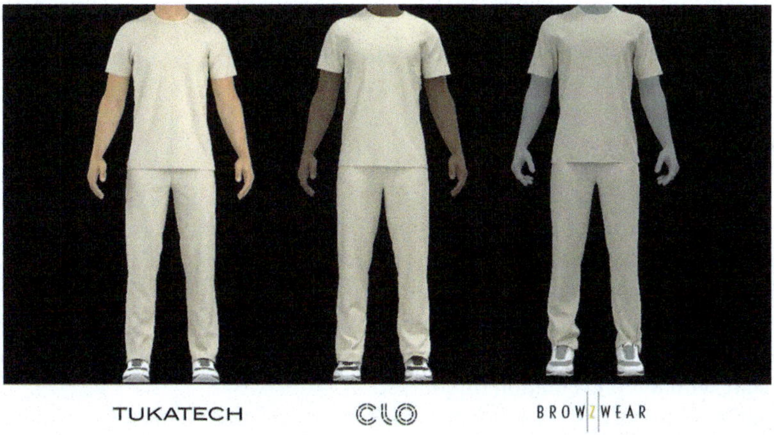

Fig. 7 TUKA3D [14]

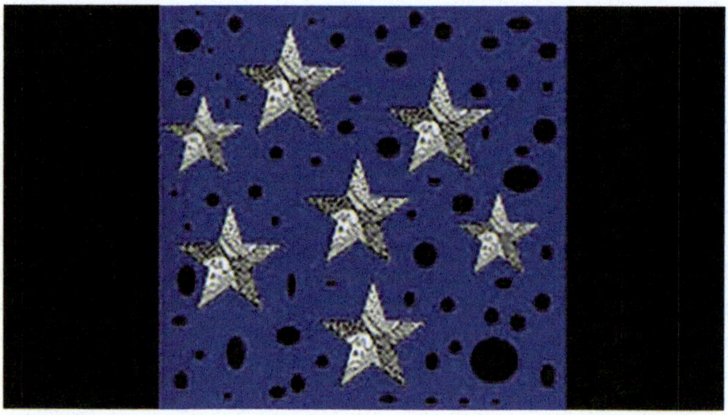

Fig. 8 Spoonflower's free design tool [13]

3.11 Seamless Studio

Anyone can make gorgeous wallpaper backgrounds and seamless textures using the Seamless Pattern Designer Android app. With just a few clicks and an intuitive editor, people can create effortless patterns. In addition to allowing you to import your own photographs for use in your project, the software comes with built-in clip art. The option to export patterns as PNG, JPEG, or WEBP pictures is one of Seamless Pattern Designer's best features. This facilitates the sharing and reuse of the designs in various other endeavours. Seamless Pattern Developer is a fantastic application to have at your fingertips whether you're a professional designer or just want to make some interesting phone backgrounds [14].

3.12 Pattern Maker

With the use of cutting-edge artificial intelligence technology, consumers can easily create original and imaginative seamless vector patterns with Pattern Maker AI in Fig. 9. Using its easy-to-use interface, Pattern Maker AI, allows you to choose from a variety of pattern types, including anime, photographic, digital art, comic book, a figment and beyond. Additionally, one can alter the number of permutations and even start with a representation of yourself [13].

After then, the strong AI generates lovely and smooth designs for you to manipulate in vector format, which you can download and utilise for product, print, and online design. Websites, mobile apps, prints, apparel, fabric, wrapping paper, scrapbooking, crafts, home décor, packaging for goods and more may all be made with the patterns [12].

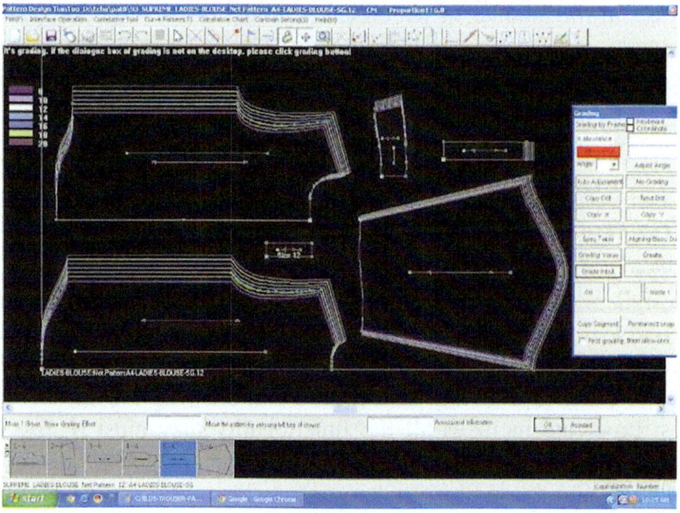

Fig. 9 Pattern Maker [12]

Fig. 10 Adobe Capture [10]

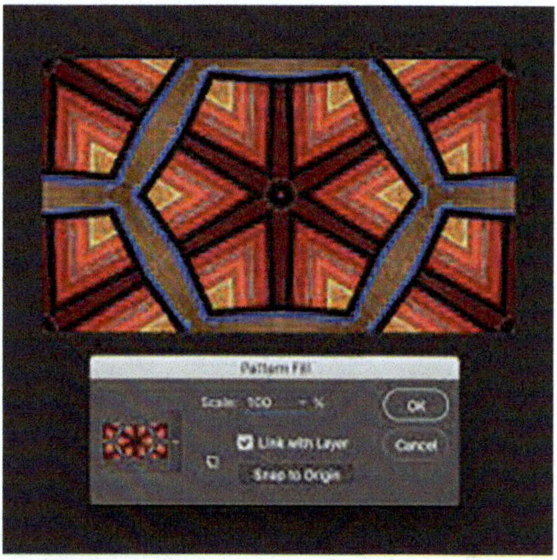

3.13 Adobe Capture

An app for your tablet or phone is called Adobe Capture. Using the camera on the gadget, can "take" colours or designs within the application, which will be transferred straight into the Files menu of the Photoshop application or Illustrator. Anyone may utilise them to decorate either Photoshop or Illustrator drawings after they're placed and saved to the Collections, as provided in Fig. 10. To discover a bit further about the manner in which Collections function, check read this wonderful explanation and overview from Adobe. This is an enjoyable method of creating original fabric designs using the hues and textures you encounter every day [10].

3.14 *Procreate*

Fashion designers wishing to produce a variety of clothing types and final applications will find Procreate to be an extremely adaptable programme. Procreate's versatility in supporting various artistic processes, from crude drawings and discussions about ideas to finished drawings and precise flat drawings, is another of its main advantages. Procreate includes all the functionality they must realise those concepts, regardless of where you are in the development phase, as emphasised in Fig. 11. Making time-lapse films to demonstrate the process of designing, guiding, and training is another great use of Procreate. A terrific approach to interact with the public is through time-lapse films, particularly whenever the goal is to show off the approach to design or encourage people through the work [4].

Fig. 11 Procreate [4]

3.15 *Digital Fashion Pro*

With the help of Digital Fashion Pro, consumers may produce designs and layouts for garments that are up to the standards of an experienced designer. It provides a variety of tools for designing clothes, such as models, materials, and surfaces, and has a simple user interface, as given in Fig. 12. Ambitious fashion designers can experiment with designs, envision principles, and realise artistic ideas in the digital realm with this tool, which is a great reference. Because it may expedite the layout stage and be a useful tool for people and startups looking to break into the fashion industry, Digital Fashion Pro is extensively used [15].

3.16 *Adobe Illustrator*

A flexible graphic design program like Adobe Illustrator may be applied to create emblems, concepts, representations, formatting, and beyond. It is widely recognised for being capable of generating content wholly adjustable, so it's essential in graphic design work and guarantees that works maintain their level of detail along with coherence regardless of magnitude. It is the global benchmark for graphic design because, in contrast to raster-based tools, it defines lines, forms, and patterns using algebraic formulas [16].

Fig. 12 Digital Fashion Pro [15]

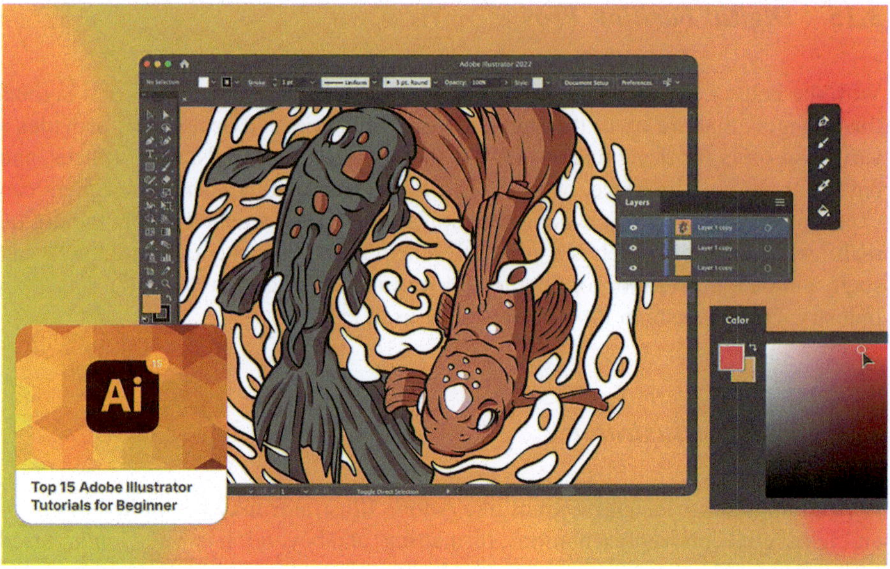

Fig. 13 Adobe Illustrator [14]

The tool panel that was layers panel, and Properties panel are some among the key components that make up Illustrator's layout. The essential instruments for creating, modifying, and operating shapes are located in the Capabilities menu. These include the Writing instrument, Gestalt, as well as Copy devices, among others, as shown in Fig. 13. The previous Strata menu is essential because it arranges multiple creative aspects. Excluding particular parts or adjusting visibility enables more effective management of intricate artwork. The Properties panel simplifies access to crucial parameters such as cerebrovascular accident, dirt, openness, and aligning by providing relevant alternatives according to the current item [14].

3.17 Adobe Photoshop

In order to transform their ideas into an appealing layout and a reproducible style, clothing and design professionals need Adobe Photoshop. As one of the oldest and most crucial tools for managing pixel-based graphics, Adobe Photoshop has made a name for itself. Photoshop's capability notwithstanding, reproducing work frequently necessitates recurring stages and may become laborious and stifle one's imagination, as shown in Fig. 14. At the same time, the designer needs to focus on both the central tile and the overall pattern. Modifications to the core tile's colours and shapes must be carefully disseminated to every unit in the whole pattern [16].

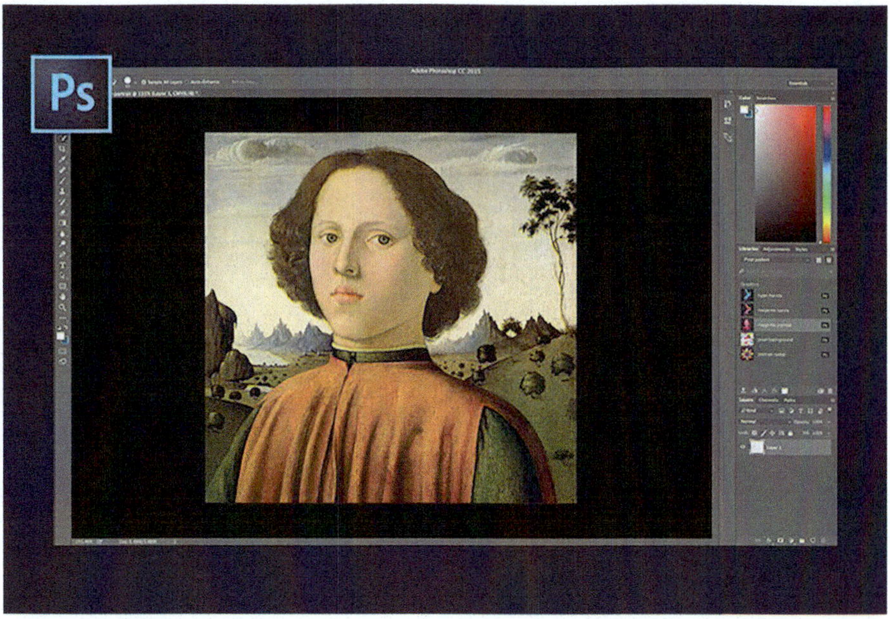

Fig. 14 Adobe Photoshop [16]

3.18 CorelDRAW

The CorelDRAW Graphics Suite offers a well-deserved reputation as an adaptable instrument for the design of graphics, but its entry into the textile design software business indicates that it has much more potential. Fundamentally, CorelDRAW provides an amazing range of functions to meet the demands of creators who want to produce intricate, custom clothing and textile designs. Designers may effectively integrate their creative visions into practical goods with CorelDRAW, as given in Fig. 15. Fashion industry professionals who want to optimise their processes through ideation to manufacturing may find this especially helpful. High customisation is possible because of the app's capacity to add patterns, textures, and colours to designs [17].

3.19 CLO 3D

CLO 3D comes next in our ranking among the most widely used apparel design tools. The perfect base for 3D apparel design and development is being provided by a powerful computer system called CLO 3D, which is reinventing the apparel industry. Recognised for its advanced additive reality capabilities, CLO 3D lets creators imagine, wrap, and assemble clothing in a multifaceted environment that is

Fig. 15 CorelDRAW [17]

Fig. 16 CLO 3D [9]

remarkably lifelike. The application speeds up the creating stage and does away with the necessity for tangible designs through its options for template creation, evaluation, and sanitation. Professionals in clothing manufacturing regularly employ CLO 3D, given that it boosts output and reduces the period from concept to the creation of apparel in a way that makes it more ecologically responsible, as shown in Fig. 16 [9].

4 Technology Limitations for Designing Textiles

Garment artists can efficiently produce, alter, and then showcase materials, shades, and patterns on computer screens by using textile creation tools. However, fabric programming has some limitations, identical to anything else, and this can prevent makers from being as imaginative and creative as they could be [18].

4.1 *Technical Expertise and Tools*

Among the main disadvantages of fabric design applications is the reality that it needs some sort of electronic know-how plus proficiency to use effectively. Artists need to learn how to use the apps, change their configuration and solve any issues that may arise. Other requirements involve possessing access to a fast and stable laptop and an excellent monitor with a suitable writer or reader. Some factors—software ignorance, lack of resources, or faraway work—can present challenges for authors [2].

4.2 *Realistic Responses Using Models*

One additional drawback of garment designing programs is their failure to faithfully mimic the texture as well as the appearance of authentic fabrics along with apparel. Artists might not be equipped to predict exactly how patterns, shades, and finishes would look and feel on different body types, in various amounts of light, or as they move. Furthermore, they could have difficulty in quickly getting feedback on the patterns from suppliers, specialists, or potential customers. Such constraints might affect the accuracy, attractiveness, and worth of the final drawings as well as individual creators' emotions of confidence and contentment [6].

4.3 *Creativity and Motivation*

Another disadvantage of cloth program design is that it may restrict artists' ability to conceive with imagination and innovatively. Dependence too heavily on pre-made plans, presets, or features of an application can restrict creators from exploring using their particular concepts and visions. Furthermore, consumer thoughts, along with assessments beyond the trends preferences, as well as attitudes of the application makers, may have an effect. These aspects might diminish the projects' authenticity, uniqueness, and variance while also adding to the artists' originality and zeal [19].

4.4 Cooperation and Correspondence

Systems for designing textiles may make it more difficult for artists to put projects together and interact with everyday pertinent parties, which is a possible disadvantage. Communicating, working together, or making changes to what works with businesses, suppliers, consumers, or other innovative people can be tricky for creators. Technological problems, which include issues with confidentiality, disputes over copyright, or challenges with integration, may additionally surface. These drawbacks may affect the efficacy, accomplishment, and total worth of the entire process of design, in addition to the relationships and confidence between the stakeholders [19].

4.5 Environmental Consciousness and Principles

The ability of cloth computer programs to overlook moral as well as environmentally conscious company procedures represents just one amongst its possible downsides. It could happen for artists to fail to consider the ways in which their products might affect culture, the environment, or local communities. Utilisation of resources, waste production, mistreatment of workers, as well as adaptation of culture are a few instances of some of these impacts. It's certainly probable people are ignorant of the effects that they intend to have on morality, ethics, or law—for example, the propagation of principles, the maintenance of works of fiction, or a respect for diversity. Such constraints could affect the creators' relevance, position and sense of responsibility in addition to the company's prospects [14].

4.6 Options and Remedies

Fabric computer programs might turn out to be a very useful tool for fashion professionals; however, it cannot entirely substitute their artistic abilities. To fully understand the endless possibilities of the system, creators ought to seek direction, counsel, or input from fellow designers, coaches, or experts regarding whether to employ it effectively. Trials may employ genuine components, attire, or dummies to get responses from a range of goods and perspectives sight. They additionally need to create their own distinct goals, designs, and characters, utilising cues from nature, history, and the arts, among other sources. On the internet, settings, devices, or systems allow consumers to engage, collaborate, or share concepts while preserving the values they value. Lastly, architects should examine and evaluate the ecological viability and integrity of their pieces to ensure they are in line with generally accepted norms. Individuals should be able to create things that aren't simply lovely but additionally morally sound and functional [3].

Fig. 17 Online presence for textile designers in current situation [20]

5 Online Presence for Textile Designers in the Current Situation

In modern times, the apparel fabric business has witnessed a dramatic shift from its historical foundation in haptic and visual stimuli. The makers of textiles need to change as society moves more and more around digital communication in order to keep and improve their exposure for viability, as given in Fig. 17. The following examines the main conclusions of current investigations on fabric creators' internet presence along with offers advice on the best approaches and resources [20].

5.1 Online Advertising

The contemporary method of using technological advances and internet-based channels to promote goods, services, and brands has been referred to as Online Advertising. Online advertising uses the potential of the web to reach a worldwide audience, in contrast with standard advertising, which depends on tangible media like print, radio, or broadcast [11].

- Create a website
- Online communities
- Networking
- Content creation [21]

5.2 Create a Website for Textile Designers

A virtual portfolio plus a formal display of their talents, knowledge, and artistic endeavours are provided by an apparel designer's webpage. Designers of textiles need to have an effective website in order to interact with customers, partners, and even potential clients in modern times. The online presence serves as a forum to showcase textile patterns, draw attention to distinctive methods, and convey the commercial identity belonging to the artist.

- Choose a domain name
- Select a website builder
- Design your website
- Customize and optimize
- Add features and functionality
- Launch and maintain [22]

5.3 Social Media

Social networks and material creation, sharing, and interaction are all part of the digital world. The social media sites TikTok, YouTube, Facebook, Instagram, Twitter, LinkedIn, and others are essential for interpersonal as well as professional interaction. Internet advertisement has turned into an efficient instrument for companies looking to increase visibility, cultivate client connections, and boost transactions using sponsored posts or natural participation.

- Define our brand
- Choose the right platforms
- Create engaging content
- Build and engage with your audience
- Optimize your profile
- Analyse and adjust
- Additional tips [9]

5.4 Online Communities

Social networks on the internet are virtual gathering places for people having similar objectives, desires, or hobbies whereby they may communicate, exchange information, plus offer to help their fellow members. Conversation panels, Facebook and Twitter groups of people, communities, and specialised connections are just a few examples of the different formats that these groups of people can take. They function as online gathering places for global connections, cooperation, and dialogue [14].

- Identify relevant communities
- Participate actively
- Create valuable content
- Leverage online platforms
- Networking and collaboration
- Showcase our expertise
- Build your brand
- Monitor and adapt [4]

5.5 *Networking*

The method of establishing and fostering business connections in order to offer chances, assistance, and knowledge is known as connecting. Building relationships based on mutual benefit entails reaching out to people who have similar interests, objectives, or professions. There are many other places where networking can happen, such as industry conferences, social events, internet forums, and trade organisations.

- Create a professional portfolio website
- Utilize social media
- Engage in online communities
- Start a blog
- Collaborate and network
- Leverage online marketplaces
- Use professional networks
- Invest in quality visuals
- Keep an eye on and adjust [6]

5.6 *E-Commerce*

The acquisition and sale of products or services using a website is referred to as online trading or electronic commerce. It includes a variety of internet-based company operations, such as digital services, trade conversations, and retailing. The method in which companies and customers engage has been completely transformed by electronic commerce, which offers an extensible, available, reliable and effective medium for activities [18].

5.6.1 Etsy

A well-liked online store for special, antique, and handcrafted items is Etsy. Customised concepts, tiny quantities, and unique or limited-production items are all produced by textile producers. Build a store to exhibit our textiles with beautiful photos, thorough descriptions, and competitive prices. For enhanced advertisements, make use of Etsy's integrated monitoring and SEO tools. Interact with a market of consumers drawn to artisanal and handcrafted goods. Participate in forums, teams, and virtual events to network and gain visibility [19].

5.6.2 Spoonflower

Spoonflower specializes in custom fabric printing, wallpaper, and home decor products. Textile designers who create patterns and prints suitable for fabric and home decor applications. Upload our designs to Spoonflower's platform to sell as fabric, wallpaper, gift wrap, and other products. Offer customers the option to purchase fabric by the yard or as finished products. Connect with other designers and enthusiasts through Spoonflower's design challenges, contests, and community forums [20].

5.6.3 Print-on-Demand Platforms

Textile designers their product offerings beyond fabric to include apparel, home decor, accessories, and stationery. Upload designs once and have them printed on demand as orders come in. Platforms handle production, shipping, and customer service, allowing designers to focus on creating. Reach a worldwide audience without the hassle of managing inventory or fulfillment logistics [23].

5.6.4 Our Website (Direct Sales)

Set up our e-commerce website using platforms like other website builders. Textile designers who want full control over branding, customer experience, and sales strategy. Customize our website to reflect our brand identity. Showcase our entire product range, offer exclusive collections, and integrate a blog to engage our customers. Improve the site's optimisation for searches to draw more natural visits. To increase traffic to our website, use promotional emails, networking platforms, and other digital advertising techniques. Build direct relationships with customers, collect customer data, and create personalized marketing campaigns to increase sales and loyalty [24].

5.6.5 Tips for Success

Use professional photography to showcase our textiles effectively. Provide clear and informative product descriptions, including materials used, care instructions, and sizing information. Optimize our product listings with relevant keywords to improve visibility on search engines within each platform. Offer excellent customer service to build trust and encourage repeat purchases. Use social media, email newsletters, influencer collaborations, and paid advertising to drive traffic to our online store or marketplace listings. By leveraging these online marketplaces and e-commerce platforms, textile designers can expand their reach, increase sales opportunities, and establish a strong online presence [25].

6 Conclusion

In summary, multimedia interaction has completely changed the design of textiles by fusing conventional artistic methods with technological advances. This merging has transformed the way architects handle their work and broadened the range of potential opportunities. Increased exploration with structures, colours, and surfaces is made possible by the incorporation of interactive software as well as platforms for multimedia, giving artists previously unheard-of freedom and accuracy. Correct model demos prior to manufacturing are made possible by immediate form visualisation tools, which reduce errors and ensure that results match the maker's intent. Internet tools also help to promote teamwork by facilitating better communication and idea refinement between producers, developers, and customers. Using mobile devices to provide individualised and customisable cloth goods increases customer happiness and commitment. Online meetings and instructional resources also make use of audiovisual to offer worthwhile learning experiences that keep either new and seasoned artists up to date on the most recent developments. Software for textile design improves both fashion design abilities and textile product manufacture. Because doing so allows innovation to emerge in the individual's endeavours and labour. Consequently, the outfit gains style. The fact that utilising fabric software for design requires hardly significant design ability is only one of its main advantages. With the help of this drafting programme, any creative or design aficionado may produce polished, virtual clothing designs. Finding something that inspires your imagination and permits you to realise the fabric creations is bound to happen using the range of alternatives offered. Multimedia immersion will probably become more and more important as the industry develops, spurring creativity and expanding the frontiers of textile craftsmanship. Adopting such innovations will help you stay competitive and advance both the science and the technique of textile creation, which is bound to end in material solutions that are more changing, practical, and visually stunning [9].

References

1. Aldrich, W. (1994). *CAD in clothing and textiles*. Blackwell Science.
2. Luniya, V. (2010). *A book of textile designs*. RG Group.
3. Wilson, J. (2001). *Handbook of textile design principles, processes and practice*. Woodhead Publishing.
4. Simon, P. (2014). *Multimedia and interactive technologies for design*. Innes, Springer.
5. Irwin, K. (2016). *Surface design for fabric*. Fairchild Books.
6. Thiel, L. L. (2016). *Digital textile printing and design*. A&C Black Publishing.
7. Smith, S. J. (2018). *Textile technology and design: From fashion to interiors*. CRC Press.
8. Lee, A. L. (2017). *Introduction to digital textile design*. Fairchild Books.
9. Brown, E. W. (2020). *Advanced textile design: Innovative digital techniques*. Bloomsbury Visual Arts.
10. Joseph, P. D., & Cherian, T. P. (2013). *CAD/CAM in clothing and textiles*. Woodhead Publishing.
11. McMahon, K. (2012). *Interactive design for new media and the web*. Routledge.
12. Russell, A. (2018). *The fundamentals of digital textile printing*. Bloomsbury Visual Arts.
13. Reynolds, L. A. (2021). *The art of digital textile design*. Thames & Hudson Publishing.
14. Briggs-Goode, A., & Townsend, K. (2011). *Textile design principles, advances and applications*. Woodhead Publishing.
15. Isaac, C., & Bowles, M. (2012). *Digital textile design, kindle* (2nd rev ed.). Laurence King Publishing.
16. Hume, R. (2020). *Fashion and textile design with Photoshop and illustrator*. Fairchild Books.
17. Wilson, J. (2001). *Introduction to computer-aided textile design*. Fairchild Books.
18. Davis, M. C. (2019). *Textile design and digital technologies*. Routledge.
19. Hann, M. (2020). *Textile design products and processes*. CRC Press.
20. Gotipamul, R. L., Vaibhav, D., & Chilveri, V. (2014). *Virtual product development in textiles*. The Indian Textile Journal.
21. Tiwary, A. (2016). Know online advertising: All information about online advertising at one place
22. McNeil, P. (2012). *The designer's web handbook: What you need to know to create for the web*. HOW Books.
23. Archana, & Paul, S. (1989). Computer-aided dobby design. *Indian Journal of Textile Research, 14*, 109–112.
24. Khalid, F., Ahmad, T., & Khalid, A. (2014). Framework for automated dobby based fabric design system. *International Journal of Scientific and Engineering Research, 5*(8), 879–885.
25. Kadole, P. V., Gotipamul, R. L., Dhanabalan, V., & Saloni, S. (2013). A review of jacquard fabrics. *The Indian Textile Journal*.

Exploring Textile Design in Different Mediums

Jayapriya S. and Suresh S. N.

1 Introduction to Textile Design

Textile design is a vast form of design that merges creativity, technical skill and innovation to create textile fabrics, which are then used in various industries. Specific Repeats are Designing patterns, textures, and structures that accentuate the visual and tactile qualities of fabrics. This is why they use traditional techniques such as weaving and embroidery, but they should also be familiar with digital textile printing and fabric manipulation in order to create beautiful textiles that perform additional functions. Textiles imbue the appearance and physical character of fashion or transform an interior environment through skilfully selected fabrics. The Cloth designers take care of the whole thing, from aesthetic matters to realistic attributes like sturdiness, comfort and sustainability. Designers, therefore, need a good understanding of material science and fibre properties, including when to use these different materials for different applications. Designers also feel pushed by smart textiles and eco-conscious fabrics, as innovation in this dynamic field is increasingly bringing new technologies and sustainability into the designer's world. This is how textile design [1].

Textile designers are also assimilating smart materials, implanted with sensors and conductive fibres, mediating between art, science and craftsmanship impacts the way we look at products and surroundings [2].

Jayapriya S. (✉)
Associate Professor and Head, Department of Costume Design and Fashion, Nehru Arts and Science College, Coimbatore, Tamil Nadu, India

Suresh S. N.
Associate Professor and Dean, School of Science, Department of Biotechnology, Rathinam College of Arts and Science, Coimbatore, Tamil Nadu, India

Sadhna et al. (eds.), *The Art and Craft of Modern Textile Design*, Sustainable Textiles: Production, Processing, Manufacturing & Chemistry,
https://doi.org/10.1007/978-3-031-86797-2_13

2 Types of Textile Design

Different types, styles and techniques of textile design are available today. These techniques help to create the look, feel and performance of a fabric as a whole [3].

2.1 Hand Drawing and Painting

Traditional hand-drawing methods of sketching and designing are still followed to create initial concept art. There can be an increment in the usage of the Pencil, Ink or Paint for Sketches on Paper. These are later transferred to the fabric. The mediums mainly used are in Haute couture and custom garments. Just as apparent in textile-based art, fashion works and interior design [4].

2.2 Digital Textile Design

Designers create elaborate patterns digitally using software like Adobe Illustrator and Photoshop, with repeat and precision design. The Means of Application is that it is used widely in fashion, dếcor (both industrial and mass textile production), due to its scalability, perfect pattern and accuracy [5]. Digital textile design is transforming traditional textile manufacturing by amalgamation of new technology with creative expression, leading to the formation of complex, customizable patterns that can be effortlessly altered and replicated. Through the usage of digital tools like CAD (Computer-Aided Design) software, creators can experiment with colours, touches, and patterns in means that were earlier difficult by manual techniques. Moreover, the integration of 3D printing and additional digital creation approaches has empowered the progression of fresh textile structures, pushing the restrictions of fabric design. Digital textile design, in addition, enables sustainable practices, as it lessens waste and allows on-demand construction, curtailing overproduction and additional records. As this field continues to grow, it unlocks novel chances for designers to generate textiles through various mediums, extending after fashion and beyond [6].

2.3 Screen Printing

The method involves applying ink to the fabric through a mesh screen in order to create a design laid out on the fabric with built-up layers. The use of medium is for the design of large-scale textile patterns such as bedding, curtains or T-shirts for both fashion and interior purposes. There is a changeover of Russian artist Liubov Popova from painting to textile design. It examines how intangible art swayed the textile patterns, causative to the constructivist association and the incorporation of fine art hooked on functional design [7].

2.4 Embroidery

Embroidery is the process of stitching threads onto a piece of fabric to make detailed and textured designs. The wear in mediums is that designers use fashion as well as interior textiles, which includes cushioning and dining table linens for ornamental activities like Elaborations or Add-ons. Additionally, one can beautify the complex artwork also with adornments. There are machine learning methods to produce preview imageries like embroidered methods of user-uploaded descriptions. It can also be modified to prevailing neural style and transmission to generate previews, simplifying supervisory in embroidery design [8].

2.5 Appliqué

Applique is a technique where small patches cut from fabric are sewn onto a larger piece to create patterns or pictures. Applications of it include fashion, costume design, patchwork and quilts. Applique is also used widely in quilting and textile arts. The use of computer-aided design (CAD) to mimic traditional appliqué projects allows avoiding association with physical appliqué exertion. A lot of research highlights the way digital printing could be employed to reproduce complicated appliqué designs on textile products, improving competence and design accuracy [9].

2.6 Fabric Manipulation

Processes such as pleating, gathering and folding are used to manipulate the shape or texture of fabric. The main medium to work on is that it is used mostly in fashion design to make structural garments. In addition, it is used in home décor items like curtains and decorative pillows [10].

3 Textile Designing in Different Mediums

Textile design is a varied and creative discipline found within the wider arena of careers in textiles, each demanding its own blend of aesthetic sensibility, functionality and technical know-how [11]. Various applications of textile mediums in the fields of fashion, interior design, art and craft, as well as in industrial textiles, play a vital role in the global economy.

3.1 Fashion Industry

One of the major sectors that benefit from Textile design is the fashion industry. So designers end up employing fabric manipulation, motion-based textures and prints with the help of different kinds of embellishments to manifest their art. Fashion textile design should be both attractive and possess functional properties that make the material not only beautiful but also comfortable, durable, and sustainable. Pleating, gathering and draping are all methods of fabric manipulation used to create unique shapes on a garment. These techniques are often used by designers to create depth and dimension in clothing. For example, they give a definition but permit designing a garment that can still move [12].

Fashion is also a textile design environment where these recycled materials can be used now instead of just in traditional applications. Eco-friendly fashion—Sustainability in Fashion—Eco-friendly and sustainable fashion is also being developed by designers who are going as far as to make fabrics from organically produced cotton, bamboo and recycled fibres, advancing the message of loving Mother Earth. Fabric customization is mostly done by printing, a step that helps designers incorporate unique patterns and graphics directly into their fashion work. There are many ways for textile designers to go from traditional block printing to modern digital prints. Digital printing specifically has opened up a wealth of possibilities to produce exact, high-definition designs on fabrics, all with less waste and quicker turnarounds. Embroidery really brings another level of craft and opulence to a fashion design. Embellishment in which yarns or threads (preferably silk, wool, metal or rayon) are sewn onto fabric to create designs and textures, and colour foundation under the thread work is created by embroidery. Hand-embroidered, well-detailed and exclusive due to its tailored fabrics are common in haute couture. Various materials, patterns and textures are created using textiles [13].

3.2 Interior Design

Textiles are incorporated with an affirmed form of designing where in, it could be a live upholstery for the place, including curtains, carpets also or even treated as wall coverage. William Morris's designs have always been popular with wallpaper, fabric and interior design that are highly decorative, intricate floral and nature-filled patterns. Figure 1, sourced after Cranbrook Art Museum, delivers an outstanding example of Morris's iconic wall covering schemes. The designs typically include multifaceted, reiterating motifs that are precisely organized to generate a visually composed and consistent look. These schemes not only complement, enhance, and part the interiors but also condense Morris's vision of assimilating art into normal life. Cultural heritage also plays a vital role in drawing inspiration for art [15].

Now more than ever, modern interior designers are turning to custom-designed textiles to create distinctive living spaces that are uniquely their own. The orientation of the design, as well as the texture of the fabric, helps set up a tone plus style in an interior. Screen printing, weaving, and fabric manipulation techniques enable designers to style textiles that match the prevailing aesthetic of a room, be it minimalist, modern or classic. Textiles for upholstery must exhibit two opposing qualities—robust constructability and suitable aesthetics, rendering the parent textile design both artistic as well as scientific. Fabrics which are Heavy-duty textiles made from jacquard or chenille are often used for upholstery to provide the desired

Fig. 1 Sample of William Morris Wallpaper [14]

aesthetic without reducing longevity. Traditional black & white textile patterns, Floral, geometric or solid, are composed to tie into the furniture and tone of the room [16].

Curtains are always on the decorating list, whether one likes to make a statement or just keep a neat and fuss-free space. Colours, prints, and textures from textile designers in draperies complement or contrast a room's mood. Screen print designs are done in patches or all over, and then different fabric manipulation techniques like folding and pleating are used to add more dimension. Textile design is also used to make rugs and carpets. The process of design involves making patterns and selecting suitable materials that match the rest of the furniture. Traditionally, hand-woven rugs like Persian or Tibetan carpets are beautifully made, and in traditional carpets, the same goes for symmetry. Depending on the theme and design, patterns can be Geometric or abstract with colours specified. With a growing emphasis on sustainability in interior design, designers are turning to eco-friendly fibres that are crafted from recycled materials, sustainable wool and natural materials. The move is part of a larger trend from across the design industry, reflecting an increasing shift towards environmentally friendly housing [17].

3.3 *Art and Craft: Language of Expression in Art and Craft*

Textile artists self-identify as fibre or textile-based yet incorporate a variety of avant-garde techniques such as threading, embroidery and appliqué to engage in social-conceptual themes of cultural identity and commentary. Working with textiles as a medium has some distinct tactile qualities that are very pleasing to the process of creation, and the sculptural possibilities of these malleable materials offer infinite imaginative expression. The art of weaving is one of the oldest and most admired textile techniques. It is accomplished by using a weave to create intricate patterns from threads or yarn. Many contemporary artists marry conventional weaving methods with modern textiles and innovate material in ways that challenge typical ideologies about what you can do with fabric [18].

Figure 2 shows the preparatory drawing by Anni Albers. Weaving has been on a newfound high since artists like Anni Albers broke the stigma around it, portraying it as an exceptional and serious form of art.

Once a purely decorative art, embroidery is now recognized as another high craft technique that fits just as well into contemporary textile art. It is still used in art to make intricate and often story-driven pieces. Appliqué allows us to have multi-layered, textured and reveal of impossible complexity done in fabric. These two meet in mixed-media art, where painting or sculpture mixes with textiles. Installation art textiles have also been used for large-scale installations. The fabric is chosen because of its flexible, versatile ways to enter into immersive spaces or sculptural figures. Because textiles are touchable, many installation artists utilize them to direct conversations about identity, memory or empathy—subjects akin to textiles

Fig. 2 Preparatory drawings for tapestries by Anni Albers, 1926 and 1927 [19]

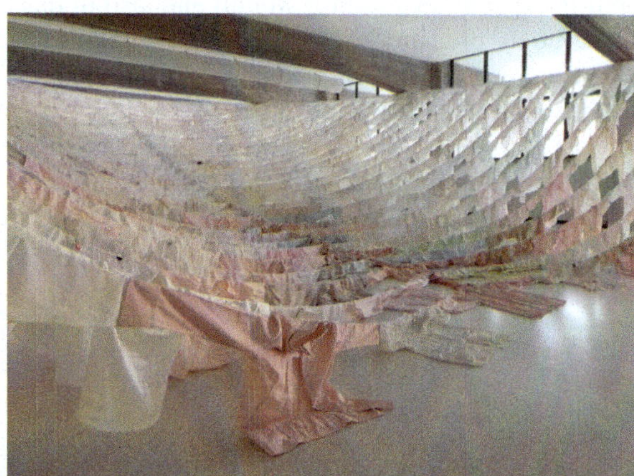

Fig. 3 Installation view of 2012 [20]

themselves that draw the viewer in through the intimacy of texture. One example is the Kaikkonen show, which works on installation designs and has proved its elegance. Figure 3 shows the installation view, which carries the perception and theme swiftly [19].

The fibre art movement of the 1960s saw artists like Sheila Hicks and Lenore Tawney create monumental textile sculptures that challenged the traditional boundary between fine art and craft. Public art commissions frequently include both murals and tapestries.

3.4 Industrial and Technical Textiles

In these types of textiles appearance becomes much less important and performance takes the front row seat [21]. The engineered textiles are designed to maintain the mesh properties (including strength, durability and environmental resistance) of traditional techniques. They can be used in industries ranging from health care to construction, military, and sports. Medical textiles are constructed with features to support patient well-being and hygiene. This includes antimicrobial fabrics for hospital gowns, bed linens, and compression garments for therapeutic benefits. Medical technical textiles need to conform to the strictest health and safety standards, which means that designers usually work toward creating materials that are both functional and comfortable for extended periods of wear. These technical textiles play a key role in creating protective clothing, including fire-retardant fabrics, bulletproof vests and hazmat suits [22].

For Example, in Fig. 4, the company Arville explains that these fabrics are woven with high-performance qualities that provide resistance to extreme heat, chemicals or impact. The same Kevlar used in bulletproof vests is designed to carry a lot of force without weight or as equally stiff material. A small but fast-growing subsection of technical textiles is "smart textiles" or "e-textiles." These materials have built-in capabilities to sense stimuli and also to react to different elements from a wide variety of sources. They could be used to monitor heart rates, regulate body temperature or even measure movement, and they have potential applications in sportswear and healthcare. A textile worn as a garment with machine-knitted biosensors embedded in the fibres is likely to be the future of wearable design and holds significant promise for everything from fitness tracking to important health monitoring applications.

Fig. 4 Ballistic fabric (Courtesy: Arville) [23]

Technical textiles have also arrived at the core of sustainability. Medical textiles, such as wound dressing and surgical gowns, are designed to satisfy specific needs for sterilization: durability and comfort. Textile designer James Pleasant designs protective clothing for firefighters or industrial workers, following safety and functionality principles practiced within textile design. Innovations are leading to the production of high-performance fabrics either by recycling old ones or using the eco-friendly process. This is particularly prevalent in sectors like athletic apparel and outdoor equipment, where consumers have demonstrated that they really care about the products and how eco-friendly they are with or without functionality [24].

4 Latest Trends in Textile Design

The evolution of textile design has never yet come to a halt; new techniques and technologies are always reforming the roots of the industry. Textile designers are experimenting with combining traditional craftsmanship with contemporary innovation to design clothes that not only look good but also stay sustainable and up-to-date. The subsequent sections explore three of the most important domains for innovation: sustainable textiles, digital printing and a combination of traditional and modern techniques [3].

4.1 Sustainable Textiles

With a strong emphasis on eco-friendly design, sustainable materials and production processes have seen increasing innovations. Organic fibres, recycled materials, natural dyeing, and unconventional woven fabrics are what textile designers are currently experimenting with. 'Sustainability' is now on the tips of every textile designer's tongue, brought about by ever-increasing consumer demand for eco-friendly products and the pressing environmental awareness issues that we are grappling with. The fashion and textile industries have come under considerable fire for their negative environmental effects, making it an area where there has been a growing effort to be as sustainable as possible at every step of the design process. Sustainable textiles are organic fibres, recycled materials, and traditional dyeing practices [25].

Designers are always pushing boundaries; instead of using basic woven fabrics they have started experimenting with an unorthodox range of woven textiles by including different fibres and weaving techniques to start taking care of the production ecological footprint. GOTS certification ensures organic fibres are taken care of compliantly throughout every stage of production, excluding heavy metal preparation dyes and unsafe peroxide that damages the colour to be more sustainable compared to conventional fibres. Sustainable The minimalist impact of organic cotton, hemp and bamboo has made them popular with textile designers. Fibers are biodegradable and utilize fewer resources like water and energy during their growth and processing. Bamboo is an example of a rapidly renewable resource as it grows really fast and without the need to use any chemicals, being one of the favourite choices for eco-friendly textiles. The role of Recycling in Sustainable Textile Design has become very important. Alternatively, designers are now using recycled fibres that are made from post-consumer or post-industrial waste, such as plastic bottles, old clothes, or fabric cuts. Using recycled material reduces the demand for virgin materials and diverts waste from landfills. One of the most commonly used eco-friendly materials in textiles is recycled polyester due to its resistance and adaptability as regular polyester but with a much lower ecological footprint [26].

Technological advancements also enable us to create top-quality fabrics from recycled fibres—providing options for designers who want performance and aesthetic materials. Natural dye use is gaining popularity in addition to the influx of sustainable fibres aimed at reducing the negative impact that comes with traditional chemical dyes. Natural dyes come from plants, minerals and other natural sources without toxic chemical fixatives. This not only prevents water pollution but also ensures that the fabric produced is non-toxic and safe for the skin. Natural dye sources including indigo, madder root and turmeric are being used to experiment with producing vivid, longer-lasting colours. Unconventional woven fabrics are the materials that are aimed at sustainable textile design. Sustainability in textiles also tasks the practices normal with weaving techniques and filmmaking. One of the areas that designers are tapping to create environmental change is weaving fabrics with biodegradable polymers and fibres made from plants that, over time, break down into harmless residue. Many of these experimental fabrics are unique in their textures and properties, allowing textile designers to further explore the possibilities for creativity. Pineapple leaves or orange peels, to give an example, are turned into fibres and then translated into fabrics which can be used in a way that job produces waste and sets new fashion horizons [27].

4.2 Digital Printing

The accuracy, detail and sophistication allowed with digital printing have made it a favourite among today's fabric designers. Detailed and colourful patterns can be printed directly onto fabric with low waste using digital printing [28]. The textile design industry is being revolutionized by digital printing because this offers a much higher level of control, flexibility and all-round performance. The colour overlay and screen printer have material waste and limited colours to choose from, requiring more time for setting. However, digital printing enables the designer to reproduce intricate and fine detail patterns directly onto fabric with little waste and practically limitless colour choices. Digital printing is highly popular among contemporary textile designers as they are accurate and digitally superior. The major advantage of working with digital printing is that it can create extremely detailed, intricate patterns that are either difficult or impossible to achieve through traditional methods. Having the capability of developing intricate and multi-layered designs with thin lines, shading or texture makes fabric design, allowing for a new range of visual depth and luxuriousness [29].

Digital printing is a method that also ensures close colour matching, capturing the smallest and finest variations to get an accurate representation of things as intended by designers. This is not only because it is capable of design but also because digital printing introduces a more sustainable way to print than traditional. Digital printers print only the amount of ink that is necessary for each design, so it minimizes waste. Digital Printing eliminates extra dye, water and chemicals to protect you from the side effect of water pollution to chemical waste. On-demand design printing also helps to mitigate the overproduction of textile fabric, a common problem within the textile industry. A completely digital print process also facilitates customization and low MOQ (minimum order quantity) ready-to-wear, having a major impact on independent designers & sustainable fashion by reducing fabric wastage [30].

Vibrant Colour & Infinite Possibilities are possible. With digital printing technology, there are almost infinite colour possibilities and all those pop kinds of colours, so testing out a lot of different colours, like bold reds, can be quite different from the hundreds that you might acquire traditionally. Digital printing is versatile even in terms of the fabric it can print on: natural fibres (cotton, linen) or some synthetic fibres like polyester and nylon. This flexibility makes digital printing adaptable to a wide range of products, particularly in the fields of fashion and apparel, home textiles, and accessories. Efficiency and speed are also promising areas. One more outstanding point about digital printing is that this process is beneficial for many reasons. The printer can handle the designs as is, which saves you a lot of time with setup and preparation. Improve production response times and more flexible production schedules which is particularly helpful when catering to the high-paced rhythm of the fashion industry. It allows work to be prototyped and sampled before production without limiting the creative potential of its designers [31].

4.3 Traditional Versus Modern Techniques Fusion

Using traditional methods like hand embroidery and fabric manipulation, textile designers fuse them with contemporary technology to produce a hybrid of the two. Over the past few years, textile designers have also combined traditional hand skills with new technology for hybrid solutions that respect tradition but incorporate innovation. Digital tools and modern materials are layered with hand processes: hand embroidery, weaving, and fabric manipulation. Streams weave in timeless and state-of-art techniques influencing one another [32]. Embroidery is one of the oldest and most elaborate textile techniques, and it has a long history in different corners of cultural regions around the globe. Now, the speed and precision of digital embroidery machines is combined with the eternal patience and professionalism that goes into hand positioning; human innovation remains even in the artistic minds. To add further visual and creative interest, mechanization can go beyond the simple patterns typically produced by today's most common digital embroidery machines to achieve highly complex and elaborate patterns yet still retain a hand-stitched feeling [33].

Additionally, digital embroidery opens up the potential for mass customization so designers can create unique embroidered items in quantity. 3D printing and fabric manipulation play a vital role. Throughout history, fabric manipulation processes have been used to develop individual textures or collaged forms and shapes in fabrications. In the 3DBra project, we take these classic approaches and combine them with the precision of 3D printing to change what is possible with cloth. For instance, making embellishments out of 3D-printed materials that will be worked into a fabric can reveal sculptural looks which are eye-catching and structurally groundbreaking. This melding of handcraft and technology enables precision and intricacy which is not achievable with traditional tools alone. Weaving with New Materials is the find of the year. Another centuries-old craft that has been repurposed for the digital era is weaving. Nowadays textile designers use CAD software to display complicated woven patterns which can be easily made by an automatic loom. In this way, intricate designs and textures can be created, combining the expertise of hand weaving with the production capacity of modern machinery. Designers are also using new, innovative materials in their woven fabrics such as metal threads, biodegradable plastics and conductive fabrics that can turn nature into a fabric [34].

Culture maintenance and innovation play a vital role. It also helps to protect the intangible heritage and guarantees that innovation keeps pace with tradition. Traditional textile processes such as Indian block printing and Japanese shibori are a huge influence on designers today, combining old-world techniques with contemporary tools and materials. This ensures the survival of those ancient techniques while also exposing them to a new generation who likes how history combines with modernity [35].

5 Future Perspectives

The upcoming textile design transversely diverse mediums is composed of transformative development, ambitious by technological progressions, sustainability essentials, and the cumulative combination of interdisciplinary methods. As the worldwide textile industry progresses, creators will discover innovative resources and means to encounter developing consumer needs and discourse environmental necessities. Digital technology will play an essential role in determining the upcoming textile design [36]. The groundbreaking usage of digital tools in textile art installations concentrates on generating collaborative and responsive designs. These installations combine traditional textile methods with the help of modern digital essentials, empowering active consumer experiences. The effort highlights the way technology can convert static textiles into alive, adaptive parts of art, nurturing arrangement and storytelling [37].

Sustainability will continue to be a vital emphasis, swaying the investigation of biodegradable resources, bio-fabricated fabrics, and circular strategy performances. Designers will progressively arrange renewable assets, waste-lessening, and closed-loop structures, guaranteeing that textile-making is in line with universal environmental areas. The upcoming textile design transversely promises mediums to be active and forward-thinking, categorized by an all-in-one combination of convention, novelty, and ecological stewardship. This growing setting offers limitless possibilities for originality, functionality, and sustainable development [38]. "ColorwAI" is an outline applying Generative Adversarial Networks (GAN) and diffusion models to generate reproductive colourways for cloths, stressing the part of AI in innovative material design is an opportunity [39].

6 Conclusion

Textile design medium is a diverse field that combines art, craft and technology to create textiles for functional and fashion uses in different mediums. It includes a diverse range of skills like weaving, knitting, printing, dyeing and embroidery which helps in playing with textures, patterns and colours. Originally associated with fashion and apparel, the field has broadened to include other areas such as interior design, automotive design, healthcare applications and perhaps the most futuristic of all smart textiles for various technological purposes.

Because sustainability is a rising trend, the usage of eco-friendly materials and processes in it is crowding around. Designers also play with eco-friendly fabrics, recycled materials and technologies like 3D printing to create new options that lessen a negative impact on the environment. As technologies in digital tools advance further, the ideal customization is available at the grasp for textile designs that are not only more accessible but malleable to unique requirements. Integration of art and technology has enabled fabric designs to reinvent in today's world, thereby also redefining fabrics beyond their traditional utility by pushing us and projects for space, objects or apparel through various sectors.

References

1. Fraser, A. (2020). *Principles, advances, and applications in textile design*. Elsevier.
2. Baker, M. (2011). *Interior textiles: Fabrics for designing and living*. Fairchild Books.
3. Dufossé, L. (2024). New research trends for textiles, a bright future. *Textiles, 4*(3), 356–358.
4. Gollapalle, P., Kota, D. K., & Kumar, R. (2024). Exploring the Kalamkari through hand-painted textile art. *ShodhKosh: Journal of Visual and Performing Arts, 5*(2), 111–121.
5. Hogben, M. (2012). *From print, pattern, sew*. Chronicle Books.
6. Haug, A., & Su, J. (2020). Digital textile design: A review on its development and future trends. *International Journal of Textile Science, 9*(3), 51–58. https://doi.org/10.11648/j.ijts.20200903.13

7. Lodder, C. (2010). Liubov Popova: From painting to textile design. *Tate Papers, 14.*
8. Beg, M. A., & Yu, J. Y. (2020). *Generating embroidery patterns using image-to-image translation.* arXiv preprint arXiv:2003.02909. https://doi.org/10.48550/arXiv.2003.02909
9. Sharma, A., & Sharma, S. (2020). Simulation of Appliqué designs using CAD. *International Journal of Current Microbiology and Applied Sciences, 9*(2), 1234–1241.
10. Fiell, C., & Fiell, P. (2011). *Textile design: A global history of patterns and techniques.* Taschen.
11. Collier, B. J., Bide, M., & Tortora, P. G. (2009). *Understanding textiles.* Pearson Education.
12. Barnard, M. (2014). *Fashion as communication.* Routledge.
13. Gale, C., & Kaur, J. (2002). *The textile book.* Berg Publishers.
14. Cranbrook Art Museum. (n.d.). *Home.* Cranbrook Art Museum. https://cranbrookartmuseum.org/
15. Mpilo, D. (2024). Role of cultural heritage in luxury fashion branding in South Africa. *International Journal of Fashion and Design, 3*(4), 1–13.
16. Fiore, A. M., & Kimle, P. A. (1997). *Understanding aesthetics for the merchandising and design professional.* Fairchild Publications.
17. Quinn, B. (2013). *Textile visionaries: Innovation & sustainability in textile design.* Laurence King Publishing.
18. Albers, A. (1926–1927). *Preparatory drawings for tapestries.* Museum of Modern Art. https://www.moma.org/
19. Grapheine. (n.d.). *Preparatory drawings for tapestries by Anni Albers, 1926 and 1927.* Grapheine. https://www.grapheine.com/
20. Lasagne, D. (2012). *Installation view of 2012* [photograph]. Stir World. https://www.stirworld.com/
21. McQuaid, M. (2005). *Applied knowledge textiles, extreme: Design and performance.* Princeton Architectural Press.
22. Kadolph, S. J. (2010). *Textiles.* Pearson Education.
23. Arville. (n.d.). *Ballistic fabric* [image]. Arville Technical Textiles. https://www.arville.com
24. Apsit, L. (2019). Sustainability in textiles for interior. *Journal of Interior Design Blog.*
25. Fletcher, K. (2014). *Sustainable fashion: Textile design journeys.* Earthscan.
26. Gwilt, A. (2014). *Fashion design for living: Sustainable fashion and textiles.* Routledge.
27. Fraser, A., & Maclaren, P. (2012). Patterns of instruction: Using screencasts in the teaching of textile design. In *Proceedings of the 29th ASCILITE Conference* (pp. 316–320).
28. Smith, C. (2018). *The art of textile design: Hand weaving & digital innovation.* Routledge.
29. Smith, C. (2018). *Hand weaving & digital creation and reproduction of textiles.* Routledge.
30. Ntim, C. K., Ocran, S. P., & Acquaye, R. (2020). Digital textile printing: A new alternative to short-run textile printing in Ghana. *International Journal of Technology and Management Research, 2*(1), 60–65.
31. Badoe, W., Samadu, K., & Frimpong, C. (2015). Exploration of innovative techniques in printed textile design. *International Journal of Technology and Management Research, 2*(1), 60–65.
32. Braddock, S. E., & O'Mahony, M. (1998). *Techno textiles: Revolutionary fabrics for fashion and design.* Thames & Hudson.
33. Ma, C., Gao, C., & Wu, X. (2021). The application of Chinese minority pattern in modern design based on information visualization. *International Journal of Frontiers in Sociology, 3*(4), 35–42. https://doi.org/10.25236/IJFS.2021.030408
34. Quinn, B. (2013). *Textile avant-garde: Technological revolution and sustainability in textile design.* Laurence King Publishing.
35. Rossbach, E. (1980). *The art of fabric collage: An innovative approach to creative textiles.* Watson-Guptill.
36. Svensson, A. (2024). Textile design events: An investigation into textile entanglements. *Journal of Textile Design Research and Practice.* https://doi.org/10.1080/20511787.2024.2392346
37. Avila, S. T. (2011). Textile tweets. *Fiber Arts, 38*(4), 20–23.

38. Pereira, M. (2020). *Textile design: Principles, advances, and applications*. Woodhead Publishing.
39. Schaerf, L., Alfarano, A., & Postma, E. (2024). *ColorwAI: Generative colorways of textiles through GAN and diffusion disentanglement*. arXiv preprint arXiv:2407.11514. https://arxiv.org/abs/2407.11514

Textile Design Education and Training

E. Mokanaasri, Sadhna, and Rajesh Kumar

1 Introduction

Textile design, which is an important segment of fashion design, is a tactile experience by which creations like fabric patterns, texture, and designs are brought to life. It involves weaving, knitting, and printing based on imaginative color combinations that combine these techniques, taking into consideration fabrics, as well as basic aesthetic design principles. This skill is important for anyone who wants to be a textile designer, as it helps them turn conceptual ideas into material form [1, 2]. The significance of textile design, education, and training lies in its role in first equipping the student for the field. Other objectives of training fundamental to design creation are orientation with principles of design, fabric types, and structural techniques, leading to a better grasp of the core elements of the field [3].

Furthermore, it has become clear that formal education is critical for textile design, especially in the initial stages, where sufficient attention should be paid to ensure the most impactful career-founding experience possible. Bachelor programs typically provide a well-rounded curriculum that includes both imaginative and practical aspects of art [4]. Significant learning occurs during undergraduate education, including design techniques, design software, and knowledge of fabric, helping students formulate a portfolio with which to find jobs in the field. A formal education also provides access to seasoned faculty and industry connections for potential mentorship opportunities. This in turn stimulates the students' creativity to create unique designs, while also providing them a chance to work with their peers

E. Mokanaasri (✉) · Sadhna · R. Kumar
Department of Fashion Design, School of Arts and Design, Woxsen University, Hyderabad, Telangana, India
e-mail: mokanaadevreddy.98@gmail.com; dr.sadhna345@gmail.com; kumar.rajeshrk2030@gmail.com

Sadhna et al. (eds.), *The Art and Craft of Modern Textile Design*, Sustainable Textiles: Production, Processing, Manufacturing & Chemistry,
https://doi.org/10.1007/978-3-031-86797-2_14

within a structured learning environment. Last but not least, an academic background equips individuals to further refine their skills while enhancing the textile design marketplace [5].

This discussion is focused around the Career Pathway in Textile Design for the Future, which is shown in Fig. 1. It highlights the importance of textile design education and training in molding the future of the industry [6]. The demand for artists and creative individuals in this field is growing rapidly, and with an inquisitive approach to exploring the educational paths available at universities, those desiring a better grasp of this craft can easily find abundant opportunities therein. This brief should give a basic understanding of these academic programs divided by category,

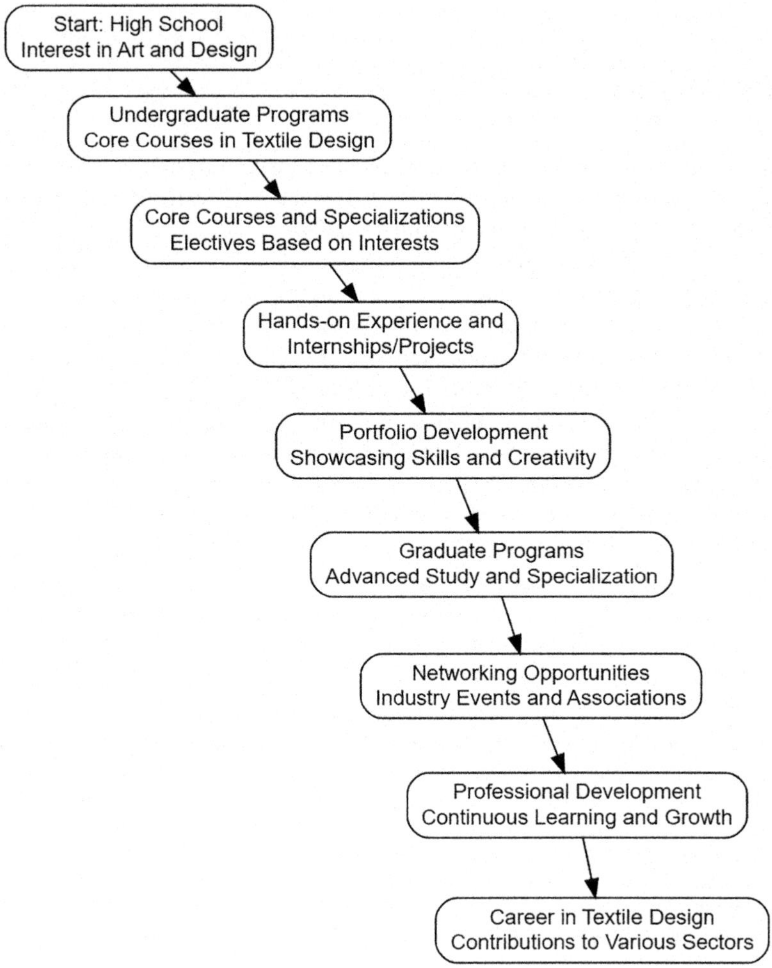

Fig. 1 Career pathway in textile design

including basics and suggestions for each, with the goal of delivering young aspirants a broader insight into the world of textile design.

2 Academic Programs and Resources

Academic programs in textile design offer a variety of avenues for students to garner the knowledge and skills needed to thrive in this creative field. They can be grouped into three primary categories: *graduate*, *undergraduate*, and *certificate* programs.

2.1 Graduate Degrees and Certificate Programs

It takes 4 years to receive a bachelor's degree in textile design, which includes comprehensive instruction on the properties of textiles, time-saving techniques, and views on style. Many of these programs also include hands-on studio classes in which students develop artistic techniques and learn to work across a range of media. Elements of Textile Design. CBSE Textbook for Class XI. The Secretary, Central Board of Secondary Education [7]. Common undergraduate degrees in textile design include [8]: Bachelor of Fine Arts (B.F.A.) in Textile Design, Bachelor of Science (B.S.) in Textile Design, and Bachelor of Arts (B.A.) in Textile Design.

There are a few graduate programs available for pupils who wish to continue their studies in textile design. Most of these programs last 1–2 years and have a specialized curriculum grounded in research [9]. Professional textile design school options include: Master of Fine Arts (M.F.A.) in Textile Design and Master of Arts (M.A.) in Textile Design. The Master's degrees in this field are highly diverse and offer numerous options according to specialization, such as digital textile printing, eco-textile design, and textile conservation, to name a few. This array allows students to personalize their education in line with their interests and professional objectives.

Most universities have certificate programs in addition to their textile design degree. These are for people who wish to learn these skills or enhance what they already know without pursuing a formal degree [10]. Certificate programs can take weeks to months to complete. A course might emphasize a specific method or subject, such as weave, screen-printing, surface pattern, digital design, or sustainability. Students interested in breaking into textile design or excelling in their field should definitely consider certificate programs. They are an accessible and quick way to learn about emerging trends in the field [11].

2.2 *Top Institutions Offering Textile Design Education*

- **National Institute of Fashion Technology (NIFT), Delhi**: One of the top institutions in India that offers Bachelor, Master, and Doctoral level programs in fashion and textile design. Its alumni are highly sought after by the textile industry given its solid academic foundation and industrial ties.
- **The National Institute of Design (NID), Ahmedabad**: Another top choice for a design institution in India that offers an interdisciplinary approach to design. Microsoft Word—NID in Ahmedabad. The school offers textile design courses with a focus on innovation and sustainability that prepares students for numerous design professions.
- **Indian Institute of Technology (IIT), Delhi**: Among the foremost internationally recognized institutions conducting programs involving engineering as well as textile design, primarily concerned with ensuring research and creativity. It is held in high esteem by both academic and corporate scholars.
- **Pearl Academy, Delhi**: Known for its practical vocational skills imparting curriculum and industry linkage. The school has a range of undergraduate and graduate fashion and textile design degrees.
- **Amity School of Fashion Technology in Noida**: Offers an in-depth course in textile design with a focus on theory as well as practical applications [12].

2.3 *Specialized Art and Design Schools*

- **Srishti Institute of Art, Design and Technology, Bangalore**: Uses a myriad of design processes to offer hands-on experience in the craft of textile design education with varied perspectives.
- **Pune's Symbiosis Institute of Design (SID)**: Offers specialized textile design degrees, emphasizing current market trends and design methods.
- **Indian Institute of Fashion & Design (IIFD), Chandigarh**: Known for its industry links and placement record, IIFD offers various textile design courses including certifications, undergraduate, and postgraduate degrees.
- **National Institute of Fashion Technology (NIFT), Mumbai**: Like its Delhi counterpart, also offers extensive textile design courses that center on real-life practices and industry-oriented learning.
- **University of Mumbai, Mumbai**: Provides textile design courses that cover different aspects and prepare students for careers in both the fashion and textile industries [12].

2.4 *Online Learning Resources and Workshops*

In textile design, online learning tools have grown significantly, since they provide prospective designers with flexible and readily available training possibilities [13]. These materials can be divided into two primary categories: webinars/workshops and MOOCs/online courses. MOOCs and other online courses can provide a vast amount of information on textile design, including modules by prestigious institutions and well-known figures in the field. Some of their best platforms and courses are as follows:

- **Fashion as Design (MoMA via Coursera)**: Concerned with the interconnected and relational field of art vis-a-vis fashion, along with textile designing, through which it tries to track the critical systems of concept development in fashion design.
- **Sustainable Fashion (Copenhagen Business School via Coursera)**: Covers the fashion and textile industry with an emphasis on current conditions in practice and theory.
- **Advanced Technologies for Textile and Fashion Industry (Texmoda)**: Focuses on the present and future technologies that influence design and manufacturing of textiles and provides the knowledge from digital sources [14].

Webinars and workshops play a vital role in the learning process for textile design, offering opportunities to gain insights from industry professionals and stay updated on current trends. These events often include presentations, demonstrations, and interactive sessions that enhance practical knowledge. For instance, webinars on digital fashion communication are highly valuable, focusing on how designers can integrate digital technology into their design and branding strategies. Similarly, textile design workshops, often conducted online, provide participants with hands-on experience in areas like textile printing, weaving, and mastering various design software. Such events not only broaden technical and creative skills but also foster professional growth and innovation in the textile design field [15].

2.5 *Industry Partnerships and Internships*

The right curriculum needs to be designed by these educational institutions in collaboration with textile companies. They assist in solving the real-world problems of the academic syllabus and provide students with enterprises that actually solve some market needs. As an example, Dr. Beenish Tahir goes on to explain the trifecta of incorporating art theory and practice in order to implement curriculums that instigate creativity and sustain the art in its continuity. Permits worldwide industry experts to work as advisors and help shape curricula, bringing industry trends and emerging technologies in the textile design field into focus [16].

There are many positions that offer scholarships and grants to support students wishing to study courses in textile design. These forms of financial support are often presented as industry cooperative scholarships amounting to funding streams through internships or project sponsorship [17]. Financial assistance is available that alleviates such concerns and allows learners to focus on their schooling while gaining hands-on experience without worrying about bills. A few colleges do the same and provide this information on their official website or orientation to make sure that they know where to apply for scholarships in order to fund their education. One such scholarship is the AATCC Foundation Textile Design Scholarship. Not just an aid, this scholarship is valid for the ones performing undergraduate students in textile design or its provided subjects having $2000. If so desired, someone can challenge themselves and pick up some work that will make their CV more competitive. This award encourages students to keep educating themselves through internships and projects with working professionals so that expertise can be demonstrated early on in their employment opportunities [18].

An internship assists textile design graduates to put what they have learned in the class into practice, helping them to reach their goals and objectives. Relating theory with practical application is encouraged; hence, an internship in tandem with their studies will provide them practical exposure to real working conditions. This helps develop crucial soft skills in each field of study, like teamwork and communication issues, which are truly important for the textile business [19]. Also, internships help to procure jobs post-education through professional referrals from individuals in the organizations they work with. A harmonious parallel between theoretical and practical training is needed to provide a textile design student with opportunities for employment growth, as it affords them tangible experience facing the real challenges of a design artist [20].

3 Overview of Textile Design Courses

Textile design courses provide a comprehensive education in the creative and technical aspects of fabric and textile production. These programs are typically structured into various levels, including undergraduate, graduate, and certificate courses, each offering specific subjects and skills essential for success in the industry.

Fiber science courses are focused on properties and classification of fibers, their chemical nature, and physical characteristics. As an example, North Carolina State University has a course named Introduction to Fiber Science, or TMS 211, and it begins by providing the students with general principles of fiber formation and structure-property relationships for polymers. Since different fibers have fixed moisture-management properties, as well as thermal, optical, and mechanical properties presently crucial to textile fibers [21].

Understanding color is essential in textile design, as it forms the basis designers follow to develop patterns and choose materials. The psychological impact of color is commonly covered in their studies, and it includes a focus on the mixture and use of color in design. Pearl Academy is an example of an institution that has integrated this theory, as it considers color theory a fundamental element in developing appealing textile designs. From the available designs, students learn how much easier it is to appeal to the target market and to produce items that are in demand [22, 23].

The practicalities of producing fabrics are generally taught in weaving and knitting courses. Students are allowed to participate in sessions where they play with techniques of textile design development, and they get to use different apparati in various experiments to evoke some beautiful new designs. Once students learn how a loom is operated, stitches are formed, and fabric structure is achieved, then and only then can they understand the complete image of textile manufacturing [24].

4 Tips for Aspiring Design Students

Aspiring design students can enhance their skills and prepare for a successful career by focusing on several key areas. The following section gives some essential tips for those looking to thrive in the design field.

4.1 Choosing the Right Program and a Strong Foundation

When selecting an institution, students should ensure it is in a good location, near textile industries and schools of design, so students can easily access job opportunities and internships. The curriculum is also crucial—opt for programs that focus on practice, with workshops and internships. They should cover critical topics like fiber science, color studies, weaving and knitting techniques, and digital media design. Also, faculty qualifications and experience are crucial; knowledgeable and skilled teachers can provide valuable guidance and mentorship, thereby ensuring comprehensive and enriching learning experiences [25]. Consider the reputation of the institution, opportunities for its alumni, and industry connections. Resources such as the Textile Skills Center and other learning outcomes websites can describe the strengths of each program.

Art and design basics are the essence of textile design, because a good grasp of visual communication, materiality, and design principles enables students to properly manage specific products or projects. Creative exploration is just as important. Experimenting with diverse media and processes helps discover the students' unique niche in design, therefore fostering innovation in areas of textile design that are conventionally defined. Building a portfolio is an essential step for aspiring designers to show off their creativity and technical skills [26]. A good portfolio should have different types of creations, including surface patterns, woven textiles, knitted designs, and process work such as sketches, mood boards, and prototypes to display the design journey. The ability to demonstrate technical skills in using CAD software is also important. To present their work well, students should focus on quality rather than quantity and ensure that their portfolio is professionally arranged and targeted at specific job opportunities or interviews [27]. This comprehensive approach reflects a designer's versatility and positions them competitively in the industry.

4.2 Networking and Professional Development

Active involvement with industry networks and communities is essential to becoming a professional in textile design. Such involvement may include membership in organizations like the Textile Society or International Textile and Apparel Association for networking, accessing resources, and receiving invitations to attend trade fairs. Attending industry events, exhibitions, and trade shows further opens up a designer's scope by helping them connect with industry players, display their work, and keep updated on trends. Building relationships with professionals and alumni is also important in career growth as it could lead to mentorship and job placement opportunities [28]. Networks such as LinkedIn are extremely helpful in connecting with experts and participating in relevant groups, thus strengthening the professional network in the textile design industry.

Developing professional networks through the textile design field is very important to professional development. A mentor can be instrumental in opening one's eyes to opportunities in employment, a portfolio, and market reentry, by providing direction appropriate to the industry's needs. Equally, competing in design competitions offers both exposure and credibility, as well as avenues for constructive critique of a new designer's work [29]. Many competitions offer opportunities to network with industry experts, including panels of judges and other professionals, thus opening avenues for future collaborations and career advancement in textile design.

Staying up to date with trends and innovations in textile design requires constant engagement with influential designers, brands, and design resources. Following popular designers and brands and reading design blogs and magazines related to textile design can help provide a constant overview of the latest trends and breakthroughs in the arena. Moreover, engagement with online forums and social media groups about textile design promotes connections, support, and access to current industry developments. This combination of learning from industry leaders and connecting with peers helps create a well-rounded understanding of the field, enabling designers to remain competitive in their careers [30].

5 Interviews with Educators in Textile Design

Prominent scholars and practitioners in textile design have reported changes defining the current trends in the discipline. A paper by the Textiles Skills Center reveals that there are changes in identifying textile education, which is more within art and design than design and technology (D&T). This transition is due to awareness of the aesthetic and social value of textiles, as well as the desire to increase students' interaction in their learning activities more creatively. Teachers have elaborated that such change has become critical in increasing students' focus and understanding of the importance of textiles in modern techniques [31].

Further interviews with educators from Moholy-Nagy University of Art and Design (MOME) have identified how students have highlighted sustainability as an essential factor. They primarily focus on post-consumer recycled material, biodegradable material, and circular economy in production. The educators found that the complexity arising from recent global events, such as the COVID-19 pandemic and environmental degradation, has contributed to students' social and ecological consciousness, which impacts the philosophies and approaches to their designs [32].

6 Conclusion

Education's role is to prepare future textile designers by guiding them with the significant knowledge they require entering this dynamically growing industry. Textile design is evolving towards sustainability, technology, and innovation, and educational programs must respond to these trends. Students in the field need to be equipped and challenged to think creatively by embracing the principles of art and design and coming up with responsible aesthetics of textile designs in order to effect a total overhaul of the fashion and textile industries. Students must be well informed about academic programs, facilities, and institutions, as this will assist them in making the right decisions regarding their education. This chapter emphasizes the facilitation of changes in textile design education that will help designers with challenges and opportunities in the future.

References

1. Tapkir, S. (2023, August 25). *A comprehensive guide to textile design courses: Exploring your options—Best designing institute in Pune*. Best Designing Institute in Pune. https://insdindia.com/a-comprehensive-guide-to-textile-design-courses-exploring-your-options/
2. FibreGuard. (2023, September 19). *Textile design: The essential introduction for beginners*. https://fibreguard.com/sell/blog/textile-design-the-essential-introduction-for-beginners
3. Human Ecology and Family Sciences. (2024). Design for fabric and apparel. In *Human ecology and family sciences* (pp. 153–158). NCERT.

4. Meyer, M. W., & Norman, D. (2020). Changing design education for the 21st century. *She Ji: The Journal of Design, Economics, and Innovation, 6*(1), 13–49. https://doi.org/10.1016/j.sheji.2019.12.002
5. Joshi, V., Saha, B., Gupta, S., Gera, P. K., Jha, B., Mathur, K., Rana, S. S., Jain, A., & Choudhary, K. (2013). Elements of textile design. In *CBSE textbook for class XI*. The Secretary, Central Board of Secondary Education.
6. Khan, M. M. (2024). *From threads to trends: Exploring a career in textile design*. Mentoria. https://mentoriablog.azurewebsites.net/from-threads-to-trends-exploring-a-career-in-textile-design/
7. Central Board of Secondary Education & National Institute of Fashion Technology. (2013). Elements of textile design: Text book class XI. Central Board of Secondary Education. https://cbseacademic.nic.in/web_material/publication/cbse/34ElementTextileDesigningXI.pdf
8. Shrivastava, R. (2024, November 7). *Textile design course, eligibility and career opportunities*. Krmangalam. https://www.krmangalam.edu.in/textile-design-course-eligibility-and-career-opportunities/
9. Mohapatra, L. (2024, March 13). *The scope of Indian fashion education: Exploring an alternative model*. Fibre2Fashion. https://www.fibre2fashion.com/industry-article/9928/the-scope-of-indian-fashion-education-exploring-an-alternative-model
10. Bisht, A. (2024). *Textile design syllabus 2024: Subjects, semester-wise syllabus PDF, top colleges, books*. https://www.shiksha.com/design/textile-design-syllabus-chp.
11. Ray, I. (2024, April 3). *Textile designing courses: Details, eligibility, fees, admission, top colleges, syllabus, scope 2024*. Collegedunia. https://collegedunia.com/courses/textile-designing
12. Textile Engineering. (2023, February 12). *Top 10 textile designing colleges in India*. https://textileengineering.net/top-10-textile-designing-colleges-in-india/
13. Radulescu, I. R., Dinis, A., Malengier, B., Cupar, A., Blaga, M., & Polansky, R. (2022). E-learning course of software for textile design. In *Proceedings of the International Conferences on E-Society 2022 and Mobile Learning 2022* (pp. 258–262). https://doi.org/10.33965/es_ml2022_202202c035
14. Chowdhury, R. (2023). Digital learning in textile design: The rise of MOOCs and virtual workshops. *International Journal of Fashion Design, Technology and Education, 16*(2), 145–158.
15. Johnson, A. (2022). Innovative online learning platforms for textile design: A review of MOOCs and specialized webinars. *Textile Design Journal, 10*(3), 233–247.
16. Tahir, B. (2022). Textile design curriculum: A collaboration between textile education and industry. *International Journal of Innovation, Creativity and Change, 16*(1), 381–383.
17. Banerjee, P. (2024, November 23). *Corporate scholarships for Indian students and opportunities in 2025*. https://www.linkedin.com/pulse/corporate-scholarships-indian-students-opportunities-2025-banerjee-fvgef/
18. AATCC. (2023, November 6). *Textile design scholarship*. https://www.aatcc.org/scholarships/design/
19. Young, D., S., & Baker, R., E. (2004). Linking classroom theory to professional practice: The internship as a practical learning experience worthy of academic credit. *Journal of Physical Education, Recreation and Dance*. https://doi.org/10.1080/07303084.2004.10608536
20. *Importance of internships in fashion design degree programs*. (2024, July 10). https://www.amordesign.org/blog/importance-of-internships-in-fashion-design-degree-programs
21. The Textile Institute. (2020, July 16). *Physical properties of textile fibres* (4th ed). https://www.textileinstitute.org/product/physical-properties-of-textile-fibres-fourth-edition/
22. Textile Industry. (2023, June 25). *Understanding fundamentals of color theory in textile design*. https://www.textileindustry.net/color-theory-in-textile-design/
23. Aiman. (2023, August 10). *Importance of color theory for designer in textile design*. Textile Mentor—Textile Industry, Textile Designing, Fashion Industry and Role of Textiles. https://textilementor.com/importance-of-color-theory-for-designer/
24. Smith, J., & Johnson, A. (2021). The importance of weaving and knitting courses in textile design education. *Journal of Textile Design Research and Practice, 9*(2), 123–138.

25. Jaiswal, R. (2024, July 24). *Textile design courses, check admission process, eligibility, syllabus, top colleges, job profiles*. Physics Wallah. https://www.pw.live/exams/design-architecture/textile-design-courses/
26. Cornelia, B. (2024). Relation between art, design and craft: The importance of learning artistic techniques textile in the formation of future fashion designers. *Review of Artistic Education, 28*, 195–200. https://doi.org/10.35218/rae-2024-0024
27. Hameed, U. (2024). *Crafting excellence: Building an impressive textile designer portfolio*. https://doi.org/10.13140/RG.2.2.36552.70404
28. Jonker, S. (2023, October 18). *Technical textiles industry associations and links*. Textilesinside. https://textilesinside.com/technical-textiles-industry-trade-association/
29. Abbasi, S. H. (2024, October 29). *Seek mentorship: Learning from experienced professionals can offer guidance on career development and help you navigate industrychanges*.https://www.linkedin.com/pulse/seek-mentorship-learning-from-experienced-can-offer-guidance-abbasi-vegxf/
30. Gazzola, P., Pavione, E., Pezzetti, R., & Grechi, D. (2020). Trends in the fashion industry. The perception of sustainability and circular economy: A gender/generation quantitative approach. *Sustainability, 12*(7), 2809. https://doi.org/10.3390/su12072809
31. Foxall, D., & Ballam, R. (2024). *Unravelling the fabric of textiles education*. Textiles Silk Centre. https://www.dropbox.com/scl/fi/hlspver1kzm71ris4cspz/TSC-Unravelling-the-fabric-of-textileseducation.pdf?rlkey=eocbrrldsn2qo6evm491nn7np&e=2&dl=0
32. Tempus Public Foundation. (2023, September 15). *The future of textile design—Interview of Moholy-Nagy University of Art and Design, MOME*. Study in Hungary. https://studyinhungary.hu/blog/the-future-of-textile-design-interview-of-moholy-nagy-university-of-art-and-design-mome

The Conclusion: Recap of Insights

Sadhna, S. Greeshma, Rajesh Kumar, and E. Mokanaasri

1 Introduction

Textile design has evolved from simple hand-woven fabrics to the intricate designs that modern technology allows today. Parallel to some of these fundamental cultural shifts and technological advancements are transformations in the role of textiles in society. Textiles have served both practical purposes and as a medium for artistic expression and identity for centuries. Early fabrics date back to 5000 BCE in Africa and China, when materials such as silk, cotton, and hemp were used for clothing and other purposes. The establishment of trade networks further stretched the status of textiles and turned them into valuable commodities throughout Europe, the Middle East, and Asia [1].

The Industrial Revolution marks a decisive turning point in the history of textile production. The spinning jenny and the power loom were significant engineering breakthroughs that reduced the time involved and made it much easier to make textiles. Mass production was finally possible, ending the era of old-style handcrafting prevalent before this period [2]. Fourteenth-century innovations in dyeing technology opened new markets for textiles in Europe, and specialized production areas like Italy could build up a sector of damask fabrics [1].

Over the past several decades, digital technologies have revolutionized textile design, unlocking new possibilities and enhancing creativity in previously unimaginable ways. Advanced software packages have enabled designers to create the most customized and intricate patterns that could not be possible through other traditional methods. The shift from 2D to 3D design paradigms has also allowed for dynamic representation of textiles, enhancing creativity and accuracy in design. Another

Sadhna · S. Greeshma (✉) · R. Kumar · E. Mokanaasri
Department of Fashion Design, School of Arts and Design, Woxsen University, Hyderabad, Telangana, India

Sadhna et al. (eds.), *The Art and Craft of Modern Textile Design*, Sustainable Textiles: Production, Processing, Manufacturing & Chemistry,
https://doi.org/10.1007/978-3-031-86797-2_15

development of this technology is the ability to provide fast and accurate copies of a designer's idea for making intricate designs. This also removes screens and films that result in precise details and more vivid colours. New inventions have resulted in intelligent textiles with electronic sensors embedded within the fabrics. Therefore, even more applications will be found in fashion and health [3].

The current textile designer must balance the aesthetic requirements with sustainability as their primary goal. Going green with material and production is no longer a trend; it is essential for survival under the thrust of consumers and growing environmental awareness. A sense of responsibility is increasingly being brought into work, using organic materials and new, innovative techniques for production that create less waste.

Textiles also play a role in personal identification and cultural aspects. They allow people to embrace their heritage, or they find a way of expressing contemporary values. Globalization continues to take its toll as it influences ways of designing: cross-cultural experiences that enrich this textile landscape remain possible. These themes are explored in depth in the interplay between tradition and innovation in modern textile designs [4].

2 The Evolution of Textile Design

Throughout history, human ingenuity and adaptability have driven the evolution of textiles across various cultures and eras. From the simple methods used by ancient civilizations to the innovations shaped by changing social demands, the design of textiles reflects the timeline of human creativity and resourcefulness. Mechanism's initiation during the Industrial Revolution marked an essential acceleration in production levels. It also increased the scope of fabrics present in the market [1].

In the contemporary world, digital printing technology has revolutionized textile design with its rapid prototyping and customizing ability. CAD software can now produce very vibrant patterns with minute details, streamlining the creative process and minimizing material waste. Another development is breaking barriers, that is, the creation of specific fabric structures according to individual requirements through 3D printing [5].

Today's textile design follows the principles of sustainability. People's awareness of their environment is more significant than ever due to the crucial issues concerning the human world; hence, designing aims to find the most suitable eco-friendly materials and techniques. The prospects of biodegradable fabrics, recycling technologies, and so much more lead this industry to move toward a far more sustainable future. It finally boils down to finding an ideal proportion between classical artisanship and modern technological improvements in creating culturally meaningful and purposeful products [6].

3 The Role of Textile Designers

A textile designer combines creativity, technical skills, and sustainability to create functional and aesthetic fabrics. Their work includes pattern creation, material selection, and production guidance, often using CAD technology. Beyond beauty, they ensure fabrics meet durability, comfort, and environmental standards. Sustainability is central as designers choose eco-friendly materials and processes to balance quality and ecological impact. They collaborate with fashion brands and manufacturers by innovating new techniques and market-aligned trends. With the rise of conscious consumerism, designers spearhead initiatives toward responsible practices to create a more environmentally friendly industry while showcasing cultural heritage in their designs.

4 The Role of Technology in Textile Design

Technology has immensely impacted textile design over the last few decades in terms of remodelling production processes and creating the scope for creativity. From digital printing techniques that allow designers to apply designs directly onto fabrics to smart textiles embedded with electronic components, technological advancements have opened a new door for innovation [7].

Such integration has made production more efficient and enabled designers to use new materials and methods to make fashion a more sustainable and responsive industry. Technology will continue to impact textile design in fashion in terms of creativity and functionality [8].

Digital technologies make the design process easier by allowing for fast iterations and customizations with minimal material waste. For instance, digital fabric twins allow designers to simulate a fabric's properties before its actual production takes place, saving resources while enhancing efficiency.

Smart textiles are the latest innovation in the industry. Fabrics can now interact with the environment or wearer through sensors implanted to measure health parameters or react to changes in temperature. This also improves functionality and taps emerging markets based on health and wellness [9].

Besides, automation increases the quality of textile production due to streamlined procedures and minimal human errors. As environmental degradation becomes a significant concern, technology provides an avenue for sustainability through the use of proper recycling methods and biodegradable materials [10].

5 Sustainability in Textile Design

Sustainability in textile design has become a pivotal area of focus, driven by technological advancements and evolving consumer preferences toward environmentally responsible practices. Contemporary strategies emphasize the material innovation, energy-efficient manufacturing processes, and circular economy principles to reduce environmental impact while enhancing product functionality. Material Innovation constitutes fundamental pillar of sustainable textile design [11]. Biodegradable polyester textiles, such developed by Duvaltex, are engineered with bio-catalysts that facilitate anaerobic digestion in landfill environments. This design offers a dual end-of-life pathway, enabling either technical recycling or biological decomposition. By reducing dependence on conventional synthetic polymers, these materials contribute to mitigating the environmental impact of textile waste, which exceeds 92 million tons annually [12]. Production Process Optimization utilizes digital technologies to reduce resource consumption. Additive manufacturing, such as 3D printing, minimizes raw material waste by using only the exact amount required, while also enabling the creation of intricate geometries that are often unattainable through conventional manufacturing techniques. Similarly, digital textile printing significantly decreases water usage by up to 95% compared to traditional dyeing, employing water-based inks and eliminating the need for water-intensive post-processing [11]. Virtual sampling represents a significant innovation in sustainable fashion practices by replacing physical prototypes with digital counterparts, thereby reducing material waste by upto 75%. This approach is further advanced through the implementation of Enterprise Digital Twins (EDTs), which simulate entire production systems. Such simulations enable companies, including brands like Kenzo to optimize product designs prior to physical production and significantly minimize the risk of overproduction [13]. Blockchain technology contributes significantly to enhancing transparency across supply chains, thereby facilitating the ethical sourcing of materials such as organic cotton and recycled fibers. However, challenges remain concerning the end-of-life management of electronic textiles, which has led to the development of design frameworks that promote modularity to enable easier disassembly and recycling of components [14].

6 Conclusion

This is how the transformation of modern textile design, from traditional craftsmanship to contemporary innovation, merges creativity with technology and sustainability. As this chapter best illustrates, textiles are artefacts imbued with functional purposes and cultural narratives, reflecting individual identities and societal values. Emerging technologies have transformed the way fabrics are made, with the goal of mass customization and integration of eco-friendly aspects critical to dealing with our planet's environmental problems. The role of designers is key to determining this landscape as they strive to combine aesthetics with functionality and prioritize sustainability throughout the process. Considering the future of textile design, it can be easily observed that innovations will shape this dynamic field by integrating the best from the traditional world with modernity while facilitating cultural expressions through diversified designs influenced by global trends. Embracing these philosophies wholeheartedly will enable textile designers to take a more sustainable route while honouring heritage and responding to the world's challenges.

References

1. Kumar, M. (2023, February 2). *History of textile design*. Tutorials Point. https://www.tutorialspoint.com/history-of-textile-design
2. Un-Hjdm. (2024, September 30). *The evolution of textile design: From ancient techniques to modern innovations*. JD Institute of Fashion Technology. https://www.jdinstitute.edu.in/the-evolution-of-textile-design-from-ancient-techniques-to-modern-innovations/
3. Shah, D. (2023, April 18). *Impact of digital technologies on the textile industry*. Fibre2Fashion. https://www.fibre2fashion.com/industry-article/9622/impact-of-digital-technologies-on-the-textile-industry
4. Safiyya. (2024, September 9). *Fashion and Identity: Exploring personal syles*. Rock & Art Cultural Outreach. https://www.rockandart.org/fashion-and-identity-personal-syles/
5. Hossain, M. (2021, July 2). *An overview of digital textile printing technology*. Textile Focus. https://textilefocus.com/overview-digital-textile-printing-technology/
6. Harsanto, B., Primiana, I., Sarasi, V., & Satyakti, Y. (2023). Sustainability innovation in the textile industry: A systematic review. *Sustainability, 15*(2), 1549. https://doi.org/10.3390/su15021549
7. Munmun, I. A. (2023). The impact of technology on fashion design: From concept to creation. *International Journal of Advances in Engineering and Management (IJAEM), 5*(12), 21–27. https://doi.org/10.35629/5252-05122127
8. Sudalaimuthu, S., & Vadivu, N. (2007, September). *Speakable content*. Fibre2Fashion. https://www.fibre2fashion.com/industry-article/2433/technology-and-its-impacts-on-textile-industry
9. James, N. (n.d.). *The future of fabric: Trends in smart textiles*. Verified Market Reports. Retrieved December 31, 2024, from https://www.verifiedmarketreports.com/blog/top-7-trends-in-smart-textiles/
10. QodeNext. (2023, July 17). *Automation in textile industry: Impacts, advantages and disadvantages*. https://qodenext.com/blog/automation-in-textile-industry/
11. certilogo-commerce. (2022). 9 Interesting sustainable textile innovations. Certilogo-commerce. https://discover.certilogo.com/blogs/insights-blog/sustainable-textile-innovations.
12. Webmaster (2024, August 28). Recycled biodegradable polyester textiles. Facility Executive. https://facilityexecutive.com/recycled-biodegradable-polyester-textiles/
13. Mehta, A. (2023, November 6). Digital twins and sustainability: the future of fashion and apparel industry. https://www.incisiv.com/blog/digital-twins-and-sustainability-the-future-of-fashion-and-apparel-industry
14. Perera, N., Shahidi, A. M., Marasinghe, K., Kaner, J., Oliveira, C., Wickenden, R., Dias, T., & Hughes-Riley, T. (2024). Exploring sustainable approaches for electronic textile products and prototypes. *Sensors, 24*(17), 5472. https://doi.org/10.3390/s24175472

Index

Sadhna et al. (eds.), *The Art and Craft of Modern Textile Design*, Sustainable Textiles: Production, Processing, Manufacturing & Chemistry,
https://doi.org/10.1007/978-3-031-86797-2

The manufacturer's authorised representative in the EU is Springer
Nature Customer Service Centre GmbH, Europaplatz 3, 69115 Heidelberg,
Germany. If you have any concerns regarding our products, please
contact ProductSafety@springernature.com

Printed and bound by CPI Group (UK) Ltd, Croydon, CR0 4YY
30/09/2026
02225563-0003